GLENCOE MATH
YOUR COMMON CORE EDITION — CCSS

Assessment Masters

McGraw Hill Education

Bothell, WA • Chicago, IL • Columbus, OH • New York, NY

connectED.mcgraw-hill.com

 Education

Copyright © The McGraw-Hill Companies, Inc.

All rights reserved. The contents, or parts thereof, may be reproduced in print form for non-profit educational use with *Glencoe Math*, provided such reproductions bear copyright notice, but may not be reproduced in any form for any other purpose without the prior written consent of The McGraw-Hill Companies, Inc., including, but not limited to, network storage or transmission, or broadcast for distance learning.

STEM McGraw-Hill is committed to providing instructional materials in Science, Technology, Engineering, and Mathematics (STEM) that give all students a solid foundation, one that prepares them for college and careers in the 21st century.

Send all inquiries to:
McGraw-Hill Education
8787 Orion Place
Columbus, OH 43240

ISBN: 978-0-07-662328-0
MHID: 0-07-662328-9

Printed in the United States of America.

4 5 6 7 8 9 QDB 16 15 14 13 12

 *Assessment Masters*

Our mission is to provide educational resources that enable students to become the problem solvers of the 21st century and inspire them to explore careers within Science, Technology, Engineering, and Mathematics (STEM) related fields.

Teacher's Guide to Using the *Assessment Masters*

The *Assessment Masters* includes the core assessment materials needed for each chapter. The answers for these pages appear at the back of this book.

Are You Ready? Worksheets
- Use after the Are You Ready? section in the Student Edition.

Chapter Diagnostic Test
- Use to test skills needed for success in the upcoming chapter.
- Retest approaching-level students after the Are You Ready? worksheets.

Chapter Pretest
- Quick check the upcoming chapter's concepts to determine pacing.
- Use before the chapter to gauge students' skill level and to determine class grouping.

Chapter Quiz
- Reassess the concepts tested in the Mid-Chapter Check in the Student Edition.

Vocabulary Test
- Includes a list of vocabulary words and questions to assess students' knowledge of those word and can be used in conjunction with one of the Chapter Tests.

Standardized Test Practice
- Assess knowledge as student progresses through the textbook.
- Includes multiple-choice, short-response, gridded-response, and extended-response questions
- Student Recording Sheet corresponds with the Test Practice.

Extended-Response Test
- Contains performance-assessment tasks and includes a scoring rubric

Chapter Tests
- **AL** 1A-1B Approaching-level students; contains multiple-choice questions
- **OL** 2A-2B On-level students; contains both multiple-choice and free-response questions
- **BL** 3A-3B Beyond-level students; contains free-response questions
- Tests A and B are created with parallel format. Use when students are absent or for different rows.

Benchmark Tests
- Contains multiple-choice and short-response questions
- The first three tests provide quarterly evaluations.
- The last test provides a cumulative end-of-year evaluation.

CONTENTS

Chapter 1 Ratios and Proportional Reasoning

- Are You Ready? Review 1
- Are You Ready? Practice 2
- Are You Ready? Apply 3
- Diagnostic Test 4
- Pretest 5
- Chapter Quiz 6
- Vocabulary Test 7
- Standardized Test Practice 8
- Student Recording Sheet 10
- Extended Response Test 11
- Extended Response Rubric 12
- Test 1A 13
- Test 1B 15
- Test 2A 17
- Test 2B 19
- Test 3A 21
- Test 3B 23

Chapter 2 Percents

- Are You Ready? Review 25
- Are You Ready? Practice 26
- Are You Ready? Apply 27
- Diagnostic Test 28
- Pretest 29
- Chapter Quiz 30
- Vocabulary Test 31
- Standardized Test Practice 32
- Student Recording Sheet 34
- Extended Response Test 35
- Extended Response Rubric 36
- Test 1A 37
- Test 1B 39
- Test 2A 41
- Test 2B 43
- Test 3A 45
- Test 3B 47

Chapter 3 Integers

- Are You Ready? Review 49
- Are You Ready? Practice 50
- Are You Ready? Apply 51
- Diagnostic Test 52
- Pretest 53
- Chapter Quiz 54
- Vocabulary Test 55
- Standardized Test Practice 56
- Student Recording Sheet 58
- Extended Response Test 59
- Extended Response Rubric 60
- Test 1A 61
- Test 1B 63
- Test 2A 65
- Test 2B 67
- Test 3A 69
- Test 3B 71

Chapter 4 Rational Numbers

- Are You Ready? Review 73
- Are You Ready? Practice 74
- Are You Ready? Apply 75
- Diagnostic Test 76
- Pretest 77
- Chapter Quiz 78
- Vocabulary Test 79
- Standardized Test Practice 80
- Student Recording Sheet 82
- Extended Response Test 83
- Extended Response Rubric 84
- Test 1A 85
- Test 1B 87
- Test 2A 89
- Test 2B 91
- Test 3A 93
- Test 3B 95

Chapter 5 Expressions

Are You Ready? Review. 97
Are You Ready? Practice. 98
Are You Ready? Apply. 99
Diagnostic Test 100
Pretest 101
Chapter Quiz 102
Vocabulary Test 103
Standardized Test Practice. 104
Student Recording Sheet 106
Extended Response Test 107
Extended Response Rubric. 108
Test 1A 109
Test 1B 110
Test 2A 113
Test 2B 115
Test 3A 117
Test 3B 119

Chapter 6 Equations and Inequalities

Are You Ready? Review. 121
Are You Ready? Practice. 122
Are You Ready? Apply. 123
Diagnostic Test 124
Pretest 125
Chapter Quiz 126
Vocabulary Test 127
Standardized Test Practice. 128
Student Recording Sheet 130
Extended Response Test 131
Extended Response Rubric. 132
Test 1A 133
Test 1B 135
Test 2A 137
Test 2B 139
Test 3A 141
Test 3B 143

Chapter 7 Geometric Figures

Are You Ready? Review. 145
Are You Ready? Practice. 146
Are You Ready? Apply. 147
Diagnostic Test 148
Pretest 149
Chapter Quiz 150
Vocabulary Test 151
Standardized Test Practice. 152
Student Recording Sheet 154
Extended Response Test 155
Extended Response Rubric. 156
Test 1A 157
Test 1B 159
Test 2A 161
Test 2B 163
Test 3A 165
Test 3B 167

Chapter 8 Measure Figures

Are You Ready? Review. 169
Are You Ready? Practice. 170
Are You Ready? Apply. 171
Diagnostic Test 172
Pretest 173
Chapter Quiz 174
Vocabulary Test 175
Standardized Test Practice. 176
Student Recording Sheet 178
Extended Response Test 179
Extended Response Rubric. 180
Test 1A 181
Test 1B 183
Test 2A 185
Test 2B 187
Test 3A 189
Test 3B 191

Chapter 9 Probability

Are You Ready? Review	193
Are You Ready? Practice	194
Are You Ready? Apply	195
Diagnostic Test	196
Pretest	197
Chapter Quiz	198
Vocabulary Test	199
Standardized Test Practice	200
Student Recording Sheet	202
Extended Response Test	203
Extended Response Rubric	204
Test 1A	205
Test 1B	207
Test 2A	209
Test 2B	211
Test 3A	213
Test 3B	215

Chapter 10 Statistics

Are You Ready? Review	217
Are You Ready? Practice	218
Are You Ready? Apply	219
Diagnostic Test	220
Pretest	221
Chapter Quiz	222
Vocabulary Test	223
Standardized Test Practice	224
Student Recording Sheet	226
Extended Response Test	227
Extended Response Rubric	228
Test 1A	229
Test 1B	231
Test 2A	233
Test 2B	235
Test 3A	237
Test 3B	239
Benchmark Tests	**241**
Answers	**A1**

NAME _____ DATE _____ PERIOD _____

Are You Ready?

Review

> **Simplest Form**
> To write a fraction in simplest form, divide the numerator and the denominator by the greatest common factor (GCF).
> A ratio is in simplest form when the greatest common factor of the numerator and the denominator is 1.

Example 1
There are 15 adults and 55 students on a field trip. Write the ratio of adults to students as a fraction in simplest form.

$\dfrac{15}{55}$ Write the fraction.

$\dfrac{15 \div 5}{55 \div 5}$ Divide by the GCF, 5.

$\dfrac{3}{11}$ Simplify.

Example 2
There are 7 girls and 8 boys that are going on a bike ride. Write the ratio of girls to boys in simplest form.

$\dfrac{7}{8}$ Write the fraction.

$\dfrac{7}{8}$ is in simplest form since the GCF of 7 and 8 is 1.

Exercises

At the close of the day at a food stand, the following items remained: 7 hot dogs, 4 coney dogs, 3 bags of chips, 6 bags of peanuts, 2 bags of popcorn, and 12 packs of sunflower seeds. Write each ratio as a fraction in simplest form.

1. coney dogs to hot dogs   1. _____

2. bags of chips to bags of peanuts   2. _____

3. bags of popcorn to packs of sunflower seeds   3. _____

4. hot dogs to bags of peanuts   4. _____

5. bags of popcorn to hot dogs   5. _____

6. bags of chips to packs of sunflower seeds   6. _____

7. bags of peanuts to coney dogs   7. _____

8. hot dogs to bags of chips   8. _____

NAME _____ DATE _____ PERIOD _____

Are You Ready?

Practice

Write each ratio as a fraction in simplest form.

Brent's Cycle Shop	
Motorcycles	35
Mopeds	5
Four Wheelers	11
Dirt Bikes	7

1. motorcycles : mopeds

2. motorcycles : four wheelers

3. motorcycles : dirt bikes

4. mopeds : motorcycles

5. mopeds : dirt bikes

6. four wheelers : motorcycles

7. four wheelers : dirt bikes

8. dirt bikes : motorcycles

9. dirt bikes : mopeds

10. dirt bikes : four wheelers

1. _____
2. _____
3. _____
4. _____
5. _____
6. _____
7. _____
8. _____
9. _____
10. _____

Determine whether the ratios are equivalent. Explain.

11. 11 out of 14 students biked to school,
    22 out of 28 students biked to school

12. 3 out of 10 children played outside,
    4 out of 11 children played outside

13. 6 music CDs to 8 DVDs,
    18 music CDs to 24 DVDs

11. _____

12. _____

13. _____

2 Course 2 • Chapter 1 Ratios and Proportional Reasoning

Are You Ready?
Apply

1. **TRAVEL** Kimberly traveled to her friend's house and went 300 miles in 5 hours. On her way home she took a different route and traveled 420 miles in 7 hours. Are these ratios equivalent?

2. **TOMATOES** On Monday Janine picked 20 tomatoes off 4 tomato plants. On Thursday she picked 15 tomatoes off 3 tomato plants. Determine whether the ratios are equivalent.

3. **BASKETBALL** Daniel's basketball team won 23 games and lost 8 games. Write the ratio of wins to losses in simplest form.

4. **MUSIC** Mr. Jansen listened to 8 songs in 28 minutes. He later listened to 5 songs in 21 minutes. Determine whether the ratios are equivalent.

5. **MOVIE ATTENDANCE** Friday night's movie attendance is shown in the table. Write the ratio of males to females in simplest form.

Friday's Movie Attendance	
Males	58
Females	72

6. **MOVIE ATTENDANCE** Use the table in Exercise 5 to write the ratio of females to the total number of people attending Friday night's movie in simplest form.

Course 2 • Chapter 1 Ratios and Proportional Reasoning

Diagnostic Test

Write each ratio as a fraction in simplest form.

High School Enrollment	
Freshmen	125
Sophomores	130
Juniors	120
Seniors	115

1. freshmen : sophomores

2. freshmen : juniors

3. freshmen : seniors

4. sophomores : freshmen

5. sophomores : seniors

6. juniors : freshmen

7. juniors : seniors

8. seniors : freshmen

9. seniors : sophomores

10. seniors : juniors

1. _____
2. _____
3. _____
4. _____
5. _____
6. _____
7. _____
8. _____
9. _____
10. _____

Determine whether the ratios are equivalent. Explain.

11. 12 out of 36 students ate an apple,
    4 out of 12 students ate an apple

11. _____

12. 6 out of 10 bankers agree,
    7 out of 11 bankers agree

12. _____

13. 4 MP3 players to 8 cell phones,
    7 MP3 players to 14 cell phones

13. _____

Course 2 • Chapter 1 Ratios and Proportional Reasoning

Pretest

Find each unit rate. Round to the nearest hundredth if necessary.

1. 50 miles in 2 hours

2. 21 laps in 7 minutes

3. $10.50 for 3 pounds

4. Find the value of x in the pair of similar figures.

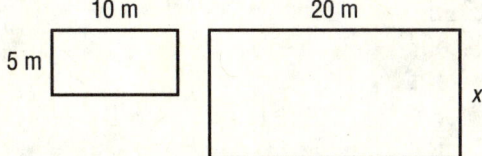

Solve each proportion.

5. $\frac{p}{7} = \frac{8}{28}$

6. $\frac{13}{26} = \frac{39}{r}$

7. $\frac{2}{3} = \frac{n}{15}$

8. $\frac{4}{x} = \frac{24}{42}$

9. **JOBS** Miley earns $10 per hour babysitting. Is the amount of money earned proportional to the number of hours she spends babysitting?

10. **FOOD** Pizzas are $10 each plus a $2 delivery fee. Is the cost proportional to the number of pizzas ordered?

1. _____
2. _____
3. _____
4. _____
5. _____
6. _____
7. _____
8. _____
9. _____
10. _____

Course 2 • Chapter 1 Ratios and Proportional Reasoning

NAME _____ DATE _____ PERIOD _____

Chapter Quiz

Find each unit rate. Round to the nearest hundredth if necessary.

1. 325 meters in 28 seconds

   1. _____

2. 128 pounds of dog food for 16 dogs

   2. _____

3. The costs of different sizes of bottled sport drink are shown. Which bottle costs the least per ounce?

Size	8 oz	16 oz	24 oz	32 oz
Price	$0.89	$1.09	$1.89	$2.39

   3. _____

Simplify.

4. $\dfrac{\frac{3}{4}}{2}$

   4. _____

5. $\dfrac{6}{\frac{2}{3}}$

   5. _____

6. $\dfrac{4}{\frac{1}{5}}$

   6. _____

7. **DRIVE** Rupert drove home at an average rate of 58 miles per hour. Find his rate in feet per second. Round to the nearest tenth.

   7. _____

Determine if the situations represent proportional relationships. Then explain your reasoning.

8. **JOBS** The table shows the amount Maggie earns each hour she babysits.

Earnings ($)	12	18	24
Time (h)	2	3	4

   8. _____

9. **SHAMPOO** The table shows the cost of shampoo at a discount store.

Cost ($)	2.95	4.50	6.05
Number of Bottles	1	2	3

   9. _____

10. Myra can fill 18 glasses with 2 containers of iced tea. How many glasses can she fill with 3 containers of tea?

    10. _____

NAME _____ DATE _____ PERIOD _____

Vocabulary Test

SCORE _____

constant of proportionality	equivalent ratios	rate
constant rate of change	linear relationship	rate of change
cross products	nonproportional	slope
derived unit	proportion	unit rate
direct variation	proportional	

Choose the correct term or phrase to complete each sentence.

1. If two quantities are (proportional, nonproportional) they have a constant ratio.

   1. _____

2. A proportion is an equation stating that two ratios are (equivalent, of different units).

   2. _____

3. Two ratios that have the same value are called (scale factors, equivalent ratios).

   3. _____

4. To determine whether two ratios form a proportion, you can find their (cross products, unit rates).

   4. _____

5. A linear relationship has (constant rate of change, variable rate of change).

   5. _____

6. The rate that describes how one quantity changes in relation to another is called the (derived unit, rate of change).

   6. _____

7. When two variable quantities have a constant ratio, their relationship is called a (direct variation, function).

   7. _____

Define each term in your own words.

8. slope

   8. _____

9. unit rate

   9. _____

Course 2 • Chapter 1 Ratios and Proportional Reasoning

7

Standardized Test Practice

Read each question. Then fill in the correct answer on the answer sheet provided by your teacher or on a sheet of paper.

1. Francesca typed 496 words in 8 minutes. Which of the following is a correct understanding of this rate?
   A. At this rate, it takes 62 minutes for Francesca to type one word.
   B. At this rate, Francesca can type 62 words in 8 minutes.
   C. At this rate, Francesca can type 62 words in one minute.
   D. At this rate, Francesca can type 8 words in one minute.

2. The table shows the prices of three boxes of cereal. Which box of cereal has the highest unit price?

Cereal Box size (ounces)	Price($)
48	5.45
32	3.95
20	3.10

   F. the 20-ounce box
   G. the 32-ounce box
   H. the 48-ounce box
   I. All three boxes have the same unit price.

3. **GRIDDED RESPONSE** A bakery sells 6 bagels for $2.99 and 4 muffins for $3.29. What is the total cost in dollars of 4 dozen bagels and 16 muffins, not including tax?

4. **SHORT RESPONSE** A teacher plans to buy 5 pencils for each student in her class. Pencils come in packages of 18 and cost $1.99 per package. What other information is needed to find the cost of the pencils?

5. During a 3-hour period, 2,292 people rode the roller coaster at an amusement park. Which proportion can be used to find x, the number of people who rode the coaster during a 12-hour period, if the rate is the same?
   A. $\frac{3}{2,292} = \frac{x}{12}$
   B. $\frac{3}{2,292} = \frac{12}{x}$
   C. $\frac{3}{x} = \frac{12}{2,292}$
   D. $\frac{x}{3} = \frac{12}{2,292}$

6. A family went on a vacation and used 5.4 gallons of gasoline to travel 150 miles. How many total gallons of gasoline will they need to travel 200 more miles?
   F. 12.6 gallons
   G. 13.1 gallons
   H. 14.3 gallons
   I. 16.2 gallons

7. **SHORT RESPONSE** You can drive your car 21.7 miles with one gallon of gasoline. At that rate, how many miles can you drive with 13.2 gallons of gasoline?

8. The speed limit on a highway is 70 miles per hour. About how fast is this in miles per minute?
   A. 4,200 mi/min
   B. 11.7 mi/min
   C. 1.17 mi/min
   D. 0.117 mi/min

9. What is the constant rate of change shown in the table?

Time (h)	Distance (mi)
0	0
1	5
2	10
3	15

F. $\frac{5 \text{ mi}}{1 \text{ h}}$ H. $\frac{10 \text{ mi}}{1 \text{ h}}$

G. $\frac{1 \text{ mi}}{5 \text{ h}}$ I. $\frac{1 \text{ h}}{2 \text{ mi}}$

10. SHORT RESPONSE At 10 A.M., the temperature was 71°F. At 3 P.M., the temperature was 86°F. Find the value of the slope and explain what it means.

11. Which of the following relationships represent a direct variation?

A.

Hours, x	1	2	3	4
Wages ($), y	20	30	40	50

B.

Hours, x	1	2	3	4
Wages ($), y	5	12	19	26

C.

Hours, x	1	2	3	4
Wages ($), y	6	12	18	24

D.

Hours, x	1	2	3	4
Wages ($), y	15	20	25	30

12. To make a punch, Anna adds 8 ounces of apple juice for every 4 ounces of orange juice. If she uses 32 ounces of apple juice, which proportion can she use to find the number of ounces of orange juice x she should add to make the punch?

F. $\frac{8}{4} = \frac{x}{32}$ H. $\frac{4}{32} = \frac{x}{8}$

G. $\frac{8}{4} = \frac{32}{x}$ I. $\frac{8}{32} = \frac{x}{4}$

13. SHORT RESPONSE A dinner is served at an athletic booster fundraiser. The constant relationship between the number of people served at dinner n and the number of ounces of beef used b is shown in the table below. How many people were served if 760 ounces of beef were used?

n	5	20	150	?
b	20	80	600	760

14. EXTENDED RESPONSE The height of the water in a bathtub is shown in the graph.

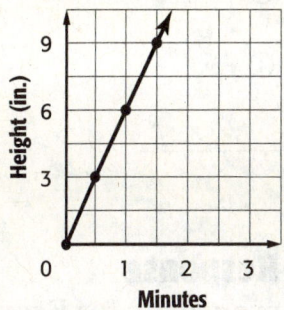

Part A Find the rate of change in inches per minute.

Part B Explain what the points (0, 0) and (1, 6) represent.

Course 2 • Chapter 1 Ratios and Proportional Reasoning

NAME _____ DATE _____ PERIOD _____

Student Recording Sheet

SCORE _____

Use this recording sheet with the Standardized Test Practice.

Fill in the correct answer. For gridded-response questions, write your answers in the boxes on the answer grid and fill in the bubbles to match your answers.

1. Ⓐ Ⓑ Ⓒ Ⓓ

2. Ⓕ Ⓖ Ⓗ Ⓘ

3. [gridded response grid]

4. _____

5. Ⓐ Ⓑ Ⓒ Ⓓ

6. Ⓕ Ⓖ Ⓗ Ⓘ

7. _____

8. Ⓐ Ⓑ Ⓒ Ⓓ

9. Ⓕ Ⓖ Ⓗ Ⓘ

10. _____

11. Ⓐ Ⓑ Ⓒ Ⓓ

12. Ⓕ Ⓖ Ⓗ Ⓘ

13. _____

Extended Response

Record your answers for Exercise 14 on the back of this paper.

10   Course 2 • **Chapter 1** Ratios and Proportional Reasoning

Extended-Response Test

Demonstrate your knowledge by giving a clear, concise solution to each problem. Be sure to include all relevant drawings and justify your answers. You may show your solutions in more than one way or investigate beyond the requirements of the problem. If necessary, record your answers on another piece of paper.

1. List three pairs of ratios that are equivalent.

2. Explain what is meant by a *proportion*.

3. **MONEY** The table below shows how much Joseph has saved the first two weeks.

Week	Amount Saved ($)
0	70
1	80
2	90

    a. Graph to find the amount Joseph has saved after 3 weeks.

    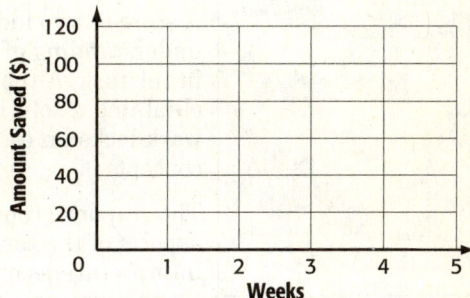

    b. What does the slope represent?

    c. Is the amount owed proportional to the number of weeks? Explain.

Course 2 • Chapter 1 Ratios and Proportional Reasoning

Extended-Response Rubric

Score	Description
4	A score of four is a response in which the student demonstrates a thorough understanding of the mathematics concepts and/or procedures embodied in the task. The student has responded correctly to the task, used mathematically sound procedures, and provided clear and complete explanations and interpretations. The response may contain minor flaws that do not detract from the demonstration of a thorough understanding.
3	A score of three is a response in which the student demonstrates an understanding of the mathematics concepts and/or procedures embodied in the task. The student's response to the task is essentially correct with the mathematical procedures used and the explanations and interpretations provided demonstrating an essential but less than thorough understanding. The response may contain minor flaws that reflect inattentive execution of mathematical procedures or indications of some misunderstanding of the underlying mathematics concepts and/or procedures.
2	A score of two indicates that the student has demonstrated only a partial understanding of the mathematics concepts and/or procedures embodied in the task. Although the student may have used the correct approach to obtaining a solution or may have provided a correct solution, the student's work lacks an essential understanding of the underlying mathematical concepts. The response contains errors related to misunderstanding important aspects of the task, misuse of mathematical procedures, or faulty interpretations of results.
1	A score of one indicates that the student has demonstrated a very limited understanding of the mathematics concepts and/or procedures embodied in the task. The student's response is incomplete and exhibits many flaws. Although the student's response has addressed some of the conditions of the task, the student reached an inadequate conclusion and/or provided reasoning that was faulty or incomplete. The response exhibits many flaws or may be incomplete.
0	A score of zero indicates that the student has provided no response at all, or a completely incorrect or uninterpretable response, or demonstrated insufficient understanding of the mathematics concepts and/or procedures embodied in the task. For example, a student may provide some work that is mathematically correct, but the work does not demonstrate even a rudimentary understanding of the primary focus of the task.

NAME _____ DATE _____ PERIOD _____

Test, Form 1A

SCORE _____

Write the letter for the correct answer in the blank at the right of each question.

1. What is the unit rate if there are 1,760 Calories in 8 servings?
   A. 176 Calories per serving   C. 228 Calories per serving
   B. 220 Calories per serving   D. 14,080 Calories per serving

   1. _____

2. A cheetah can run 70 miles per hour. What is this speed in feet per hour?
   F. 0.01 feet per hour   H. 48,400 feet per hour
   G. 210 feet per hour   I. 369,600 feet per hour

   2. _____

3. Which size of yogurt shown in the table has the lowest unit price?
   A. 6 oz   C. 10 oz
   B. 8 oz   D. 32 oz

Size (oz)	Cost ($)
6	0.89
8	1.04
10	1.69
32	4.79

   3. _____

4. Kevin can travel $22\frac{1}{2}$ miles in $\frac{1}{3}$ hour. What is his average speed in miles per hour?
   F. 65 mph   G. 67.5 mph   H. 68 mph   I. 70 mph

   4. _____

5. The table shows the cost for ordering a certain number of pizzas. What is the value of x if the cost is proportional to the number of pizzas ordered?

Pizzas Ordered	2	3	4	5
Cost	$19.98	$29.97	$39.96	x

   A. $9.99   B. $29.97   C. $49.95   D. $59.94

   5. _____

6. What is the constant of proportionality of the linear function?

Time, x	1	2	3	4
Cost ($), y	25	50	75	100

   F. 2   G. 12   H. 25   I. 50

   6. _____

7. $\frac{118}{13} = \frac{59}{z}$
   A. 1.5   B. 6.5   C. 26   D. 535.5

   7. _____

Course 2 • Chapter 1 Ratios and Proportional Reasoning

NAME _____ DATE _____ PERIOD _____

Test, Form 1A (continued)

SCORE _____

8. The table shows the relationship between time and distance. Which graph best represents the data in the table?

Speed	
Time (s)	Distance (m)
1	4
5	20
10	40

F.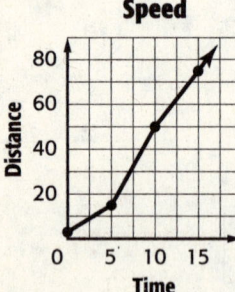

H.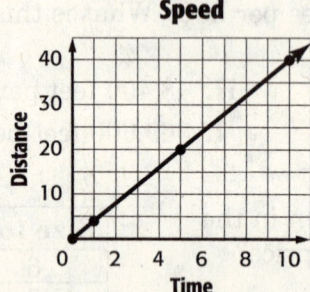

G.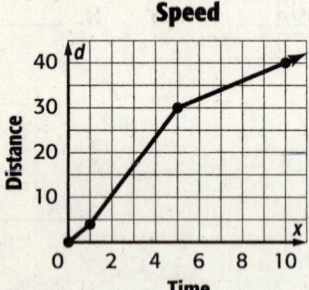

I.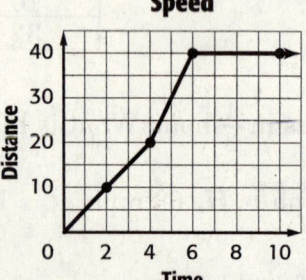

8. _____

9. What is the constant rate of change shown in the graph below?

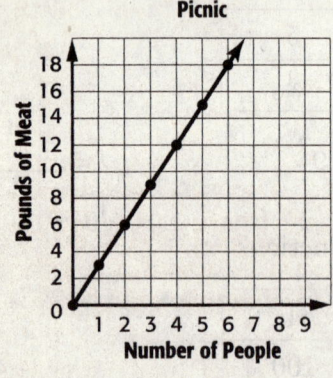

A. 2.5   B. 3   C. 7.5   D. 10

9. _____

10. What is the slope of the line from the table below?

Hour	1	2	4	8
Distance	50	100	200	400

F. 50   G. 25   H. $\frac{1}{25}$ I. $\frac{1}{50}$

10. _____

14   Course 2 • Chapter 1 Ratios and Proportional Reasoning

NAME _____ DATE _____ PERIOD _____

Test, Form 1B

SCORE _____

Write the letter for the correct answer in the blank at the right of each question.

1. What is the unit rate if there are 92 miles driven using 4 gallons of gas?
   A. 21 miles per gallon
   B. 22 miles per gallon
   C. 23 miles per gallon
   D. 96 miles per gallon

   1. _____

2. A dolphin can swim up to 40 miles per hour. What is this speed in feet per hour?
   F. 7 feet per hour
   G. 10,560 feet per hour
   H. 20,800 feet per hour
   I. 211,200 feet per hour

   2. _____

3. Which size package of pasta shown in the table has the lowest unit price?
   A. 3 oz
   B. 8 oz
   C. 16 oz
   D. 32 oz

Size (oz)	Cost ($)
3	0.99
8	2.59
16	5.69
32	11.89

   3. _____

4. Jason can travel $24\frac{3}{4}$ miles in $\frac{1}{2}$ hour. What is his average speed in miles per hour?
   F. 45 mph   G. 49.5 mph   H. 55 mph   I. 60 mph

   4. _____

5. The table shows the cost for ordering a certain number of pies. What is the value of x if the cost is proportional to the number of pies ordered?

Pies Ordered	2	3	4	5
Cost	$14.50	$21.75	$29.00	x

   A. $7.25   B. $35.50   C. $36.25   D. $43.50

   5. _____

6. What is the constant of proportionality of the linear function?

Game, x	3	4	5	6
Score, y	24	32	40	48

   F. 5   G. 8   H. 12   I. 18

   6. _____

7. $\frac{30}{42} = \frac{55}{d}$
   A. 77   B. 67   C. 39.29   D. 23

   7. _____

Course 2 • Chapter 1 Ratios and Proportional Reasoning

Test, Form 1B (continued)

8. What is the constant rate of change of the graph below?

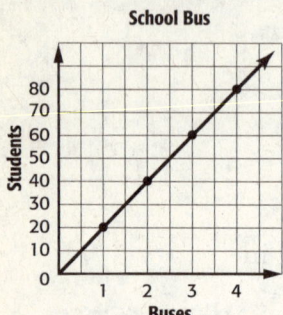

F. 20   G. 10   H. 3   I. 2

8. _____

9. What is the slope of the line from the data shown in the table below?

Time	11 A.M.	1 P.M.	3 P.M.	5 P.M.
Temperature	55	65	75	85

A. 20   B. 10   C. 5   D. $\frac{1}{5}$

9. _____

10. The table shows the relationship between the cost of an item and the length of time in months it lasts. Which graph best represents the data in the table?

Cost ($)	Length (months)
3	2
6	4
9	6
12	8

F.
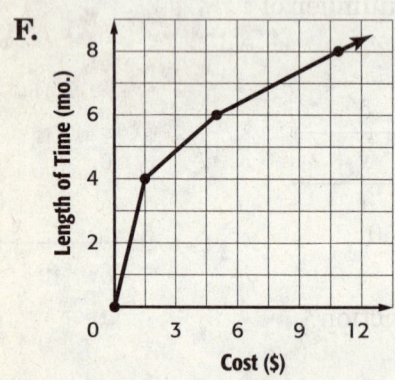

H.
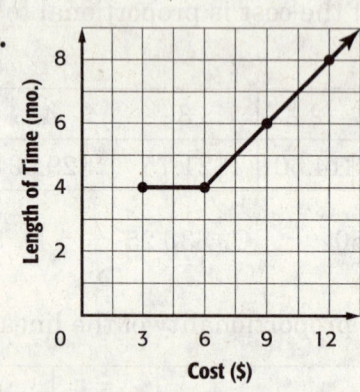

G.

I.
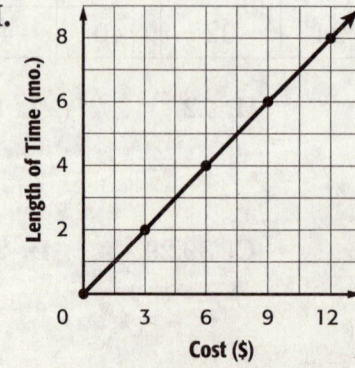

10. _____

NAME _____ DATE _____ PERIOD _____

Test, Form 2A

SCORE _____

Write the letter for the correct answer in the blank at the right of each question.

1. What is the constant rate of change of the table below?

Seconds	10	20	30	40
Meters	40	80	120	160

   A. 2 meters per second   C. 8 meters per second
   B. 4 meters per second   D. 10 meters per second

   1. _____

2. What is the slope of the line?

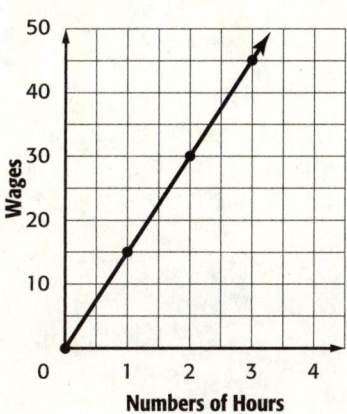

   F. 15   H. 25
   G. 18   I. 30

   2. _____

3. Selena babysits on the weekends. The equation $y = 12x$ represents the amount of money she earns. What is the constant of proportionality?

   A. 1   C. 12
   B. 6   D. 24

   3. _____

4. What is the solution of the proportion $\frac{22}{11} = \frac{r}{13}$?

   F. 26   G. 18   H. 6.5   I. 3.18

   4. _____

5. Which size can of soup shown in the table has the lowest unit price?

   A. 10 oz   C. 18 oz
   B. 15 oz   D. 32 oz

Size (oz)	Cost ($)
10	0.79
15	1.29
18	2.16
32	3.19

   5. _____

6. Cyclists were $\frac{3}{4}$ finished with their ride when they reached the 18-kilometer mark. How long was their ride?

   F. 6 km   G. 12 km   H. 18 km   I. 24 km

   6. _____

Course 2 • Chapter 1 Ratios and Proportional Reasoning   17

NAME _____ DATE _____ PERIOD _____

Test, Form 2A   (continued)   SCORE _____

7. A fox can run at a speed of 42 miles per hour. How many feet per second is this?

   A. 42 ft/s   B. 1,232 ft/s   C. 61.6 ft/s   D. 3,696 ft/s   7. _____

8. The graph shows the amount of money Joe earns each hour he works. Which statement about the graph is *not* true?

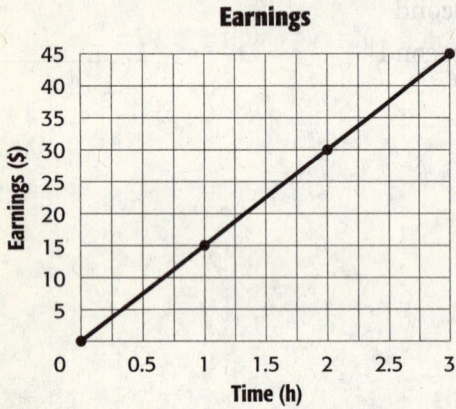

   F. The graph shows a proportional relationship.
   G. The graph shows a nonproportional relationship.
   H. The unit rate is $\frac{\$15}{\text{hour}}$.
   I. The line is straight.   8. _____

9. The table shows the cost for ordering a certain number of pieces of chicken. What is the value of x if the cost is proportional to the number of chicken pieces ordered?   9. _____

Pieces of Chicken Ordered	2	3	4	6
Cost	$3.00	$4.50	$6.00	x

10. If it takes 4 gallons of gas to drive 92 miles, how many miles can be driven using 6 gallons of gas?   10. _____

11. Martin can travel 174 miles in 3 hours. At this rate, how far can he travel in 7 hours?   11. _____

12. Bart can type 296 words in 8 minutes. At this rate, how many words can he type in 20 minutes?   12. _____

18   Course 2 • Chapter 1 Ratios and Proportional Reasoning

NAME _____ DATE _____ PERIOD _____

Test, Form 2B

SCORE _____

Write the letter for the correct answer in the blank at the right of each question.

1. What is the constant rate of change of the table below?

Hours	2	4	6	8
Miles	70	140	210	280

   A. 105 miles per hour C. 50 miles per hour
   B. 70 miles per hour   D. 35 miles per hour

   1. _____

2. What is the slope of the line?

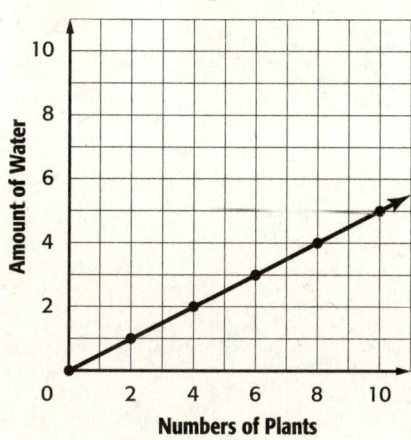

   F. $\frac{1}{2}$ G. 2   H. 5   I. 12

   2. _____

3. Dawson rakes leaves in his neighborhood. The equation $y = 10x$ represents the amount of money he earns. What is the constant of proportionality?

   A. 1   C. 10
   B. 5   D. 20

   3. _____

4. What is the solution of the proportion $\frac{3}{16} = \frac{9}{y}$?

   F. 32   G. 48   H. 60   I. 72

   4. _____

5. Which size can of green beans shown in the table has the lowest unit price?

   A. 6 oz   C. 12 oz
   B. 10 oz D. 14 oz

Size (oz)	Cost ($)
6	0.60
10	0.90
12	1.20
14	1.30

   5. _____

6. Bikers were $\frac{1}{2}$ finished with their ride at the 6-mile mark. How long was their ride?

   F. 4 miles   G. 6 miles   H. 9 miles   I. 12 miles

   6. _____

Course 2 • Chapter 1 Ratios and Proportional Reasoning

19

Test, Form 2B (continued)

7. An antelope can run at a speed of 61 miles per hour. What is this speed in yards per second? Round to the nearest hundredth.

A. 29.82 yd/s
B. 1,789.33 yd/s
C. 89.47 yd/s
D. 5,368 yd/s

7. _____

8. The graph shows the amount of money Amy earns each hour she works. Which statement about the graph is *not* true?

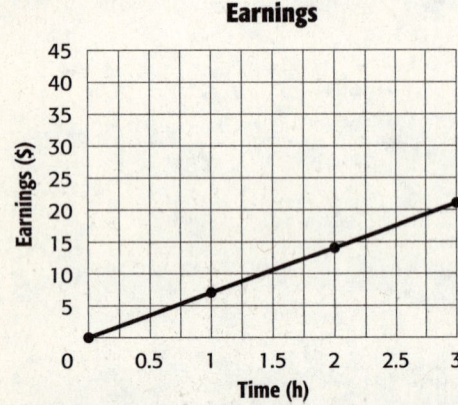

F. The graph shows a proportional relationship.
G. The graph shows a nonproportional relationship.
H. The unit rate is $\frac{\$7.50}{\text{hour}}$.
I. The line is straight.

8. _____

9. The table shows the cost for ordering a certain number of tacos. What is the value of x if the cost is proportional to the number of tacos ordered?

9. _____

Tacos Ordered	2	3	4	6
Cost	$2.60	$3.90	$5.20	x

10. If it takes 15 gallons of gas to drive 330 miles, how many miles can be driven using 20 gallons of gas?

10. _____

11. Sanjay can travel 342 miles in 6 hours. At this rate, how far can he travel in 5 hours?

11. _____

12. Patty can make 10 purses in 8 hours. At this rate, how many purses can she make in 28 hours?

12. _____

20 Course 2 • Chapter 1 Ratios and Proportional Reasoning

Test, Form 3A

1. Which size bag of cat food shown in the table has the lowest unit price?

Size (oz)	Cost ($)
24	5.49
40	8.00
64	15.99

1. _____

2. The graph shows the savings of Mia and Kyle. What does the slope of each line represent?

2. _____

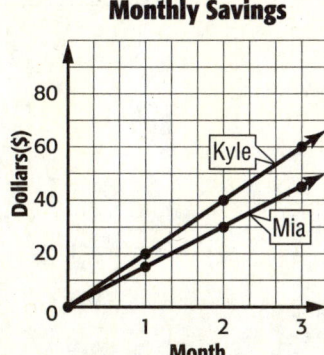

3. Cyclists were $\frac{3}{5}$ finished with their ride when they reached the 15-kilometer mark. How long was their ride?

3. _____

4. A grocery store sells 6 bottles of water for $4 and 18 bottles of water for $10. Is the cost of the water proportional to the number sold? Explain.

4. _____

Solve each proportion.

5. $\frac{k}{13} = \frac{0.6}{0.5}$

5. _____

6. $\frac{18.2}{j} = \frac{4}{16}$

6. _____

7. $\frac{6.5}{v} = \frac{3}{11.7}$

7. _____

8. When diving, a hummingbird can reach a speed of 40 miles per hour. What is this speed in feet per second? Round to the nearest hundredth if necessary.

8. _____

Course 2 • Chapter 1 Ratios and Proportional Reasoning

Test, Form 3A (continued)

9. Find the rate of change from the table.

Time (hr)	Temperature (°)
2	75
4	80
6	85

9. _____

10. Does the graph show a proportional relationship? Explain.

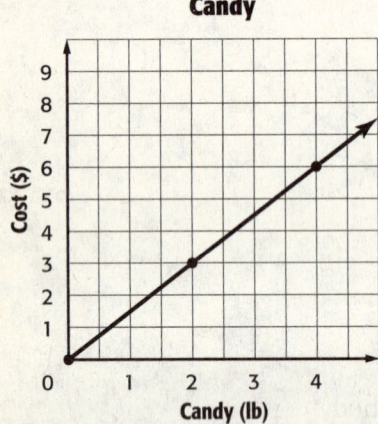

10. _____

11. If it takes the Ramirez family 3 days to travel 1,560 miles, how many days will it take them to travel 3,640 miles?

11. _____

12. Janelle can get a 24-pack of bottled water for $6.80. How much would Janelle have to pay for a 12-pack of bottled water if the ratios are proportional?

12. _____

13. Find y when $x = 9$ if y varies directly as the square of x, and $y = 245$ when $x = 7$.

13. _____

14. Jaime and Ryan work at the grocery store. The wages earned for the weekend are shown in the table and graph. Who gets paid more per hour? Explain.

Jaime's Wages	
Time (h)	Wages ($)
3	24
4	32
5	40

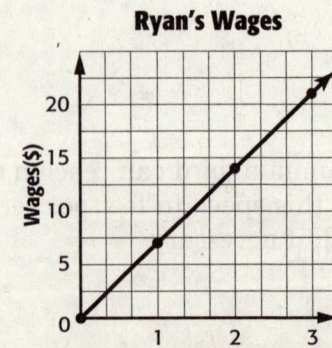

14. _____

22 Course 2 • Chapter 1 Ratios and Proportional Reasoning

Test, Form 3B

1. Which size jar of jelly shown in the table has the lowest unit price?

Size (oz)	Cost ($)
10	1.69
16	3.19
32	5.79

   1. _____

2. The graph shows the savings of Rachel and Sam. What does the slope of each line represent?

   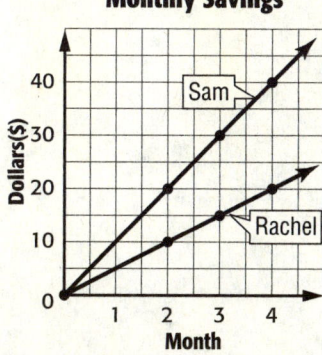

   2. _____

3. Volunteers were $\frac{3}{4}$ finished with the walkathon when they reached the $4\frac{1}{2}$-kilometer mark. How long was the walkathon?

   3. _____

4. A grocery store sells 6 bottles of water for $4.80 and 18 bottles of water for $22.50. Is the cost of the water proportional to the number sold? Explain.

   4. _____

Solve each proportion.

5. $\frac{x}{6} = \frac{3}{8}$

   5. _____

6. $\frac{4.3}{2.5} = \frac{w}{4}$

   6. _____

7. $\frac{5}{v} = \frac{2}{7.5}$

   7. _____

8. When diving, a bald eagle can reach a speed of 200 miles per hour. What is this speed in feet per second? Round to the nearest hundredth if necessary.

   8. _____

Course 2 • Chapter 1 Ratios and Proportional Reasoning

Test, Form 3B (continued)

9. Find the rate of change from the table.

Time (hr)	Temperature (°)
2	30
4	34
6	38

9. _____

10. Does the graph show a proportional relationship? Explain.

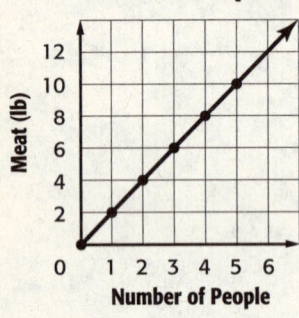

Meat Consumption

10. _____

11. If it takes the Jones family 2 days to travel 1,240 miles, how many days will it take them to travel 2,480 miles?

11. _____

12. Brian can get an 8-pack of bottled sports drink for $5.28. How much would Brian have to pay for a 6-pack of bottled sports drink if the ratios are proportional?

12. _____

13. Find y when $x = 9$ if y varies directly as the square of x, and $y = 100$ when $x = 5$.

13. _____

14. Shawn and Crystal work at the hardware store. The wages earned for the weekend are shown in the table and graph. Who gets paid more per hour? Explain.

Shawn's Wages	
Time (h)	Wages ($)
2	20
3	30
4	40

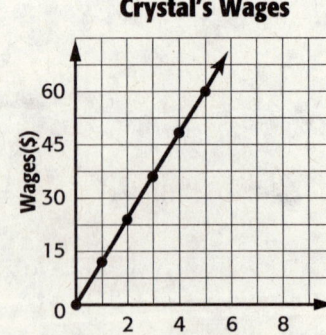

Crystal's Wages

14. _____

NAME _____ DATE _____ PERIOD _____

Are You Ready?

Review

Writing Percents as Decimals
To write a percent as a decimal, remove the percent symbol and divide by 100. When dividing by 100, move the decimal point two places to the left.

Example 1
Write 6.4% as a decimal.
6.4% = 0.064   Move the decimal point 2 places to the left and remove the percent symbol.

Example 2
Write 125% as a decimal.
125% = 1.25   Move the decimal point 2 places to the left and remove the percent symbol.

Exercises
Write each percent as a decimal.

1. 50%
2. 6%
3. 117%
4. 13%
5. 1.25%
6. 32.5%
7. 230%
8. 74%
9. 27.9%
10. 8.5%
11. 612%
12. 3.9%

Course 2 • Chapter 2 Percents

Are You Ready?

Practice

Multiply.

1. $200 \times 0.02 \times 4$

2. $30 \times 0.06 \times 3$

3. $45 \times 0.07 \times 2$

4. $350 \times 0.05 \times 6$

Write each percent as a decimal.

5. 40%

6. 8%

7. 175%

8. 80.5%

9. 9.2%

10. 432%

11. **RECREATION** On a Saturday, 55% of visitors to a zoo also went to the water park. Write this percent as a decimal.

Write each decimal as a percent.

12. 0.14

13. 0.02

14. 0.275

15. 0.076

16. 2.3

17. 5.85

18. 0.037

NAME _____ DATE _____ PERIOD _____

Are You Ready?

Apply

1. **TRACK** A track team won first place in 0.91 of the meets they competed in this year. What percent of the meets did the team win first place?

2. **FAVORITE COLOR** Mr. McGuirk surveyed the students in his math class and found that 45% of them said their favorite color was red. What decimal represents this amount?

3. **TAX** A state sales tax is 6.75%. Write this percent as a decimal.

4. **GOLF** When Akil golfs, 85% of the time he hits the green on his second shot. What decimal represents this amount?

5. **SALES TAX** The sales tax on Caden's groceries was 6.5%. Write this percent as a decimal.

6. **BANK ACCOUNT** The interest rate Calvin is earning on his bank account is 4%. Write this percent as a decimal.

Course 2 • Chapter 2 Percents

NAME _____ DATE _____ PERIOD _____

Diagnostic Test

Multiply.

1. $100 \times 0.03 \times 5$

2. $250 \times 0.05 \times 7$

3. $62 \times 0.04 \times 3$

4. $70 \times 0.08 \times 2$

5. **SCHOOL** On the last history quiz, Kanesha answered 0.75 of the 20 questions correctly. How many questions did Kanesha answer correctly?

Write each percent as a decimal.

6. 30%

7. 6%

8. 215%

9. 63.2%

10. 3.7%

11. 86.5%

12. 720%

Write each decimal as a percent.

13. 0.23

14. 0.09

15. 0.318

16. 4.28

17. 0.018

18. 0.94

19. 3.68

20. 0.171

1. _____
2. _____
3. _____
4. _____
5. _____
6. _____
7. _____
8. _____
9. _____
10. _____
11. _____
12. _____
13. _____
14. _____
15. _____
16. _____
17. _____
18. _____
19. _____
20. _____

NAME _____ DATE _____ PERIOD _____

Pretest

Find each number. Round to the nearest tenth if necessary.

1. 15% of 48

2. 25% of 80

3. 135% of 70

4. $\frac{7}{10}$% of 100

5. What percent of 50 is 36?

6. What percent of 20 is 4?

7. 30 is 50% of what number?

8. 12 is 25% of what number?

9. **COMPUTERS** A computer is on sale for $600. This is 75% of its original price. What was the original price of the computer?

1. _____
2. _____
3. _____
4. _____
5. _____
6. _____
7. _____
8. _____

9. _____

Find each percent of change. Round to the nearest whole percent if necessary. State whether the percent of change is an *increase* or a *decrease*.

10. original: 150
    new: 175

11. original: 225
    new: 180

12. original: 15
    new: 25

10. _____

11. _____

12. _____

Find the total cost to the nearest cent.

13. $9,000 car; 5% sales tax

14. $140 bicycle; 7% sales tax

15. $225 surf board; 6% sales tax

13. _____

14. _____

15. _____

Course 2 • Chapter 2 Percents

NAME _____ DATE _____ PERIOD _____

Chapter Quiz

Find each number. Round to the nearest tenth if necessary.

1. 20% of 70

2. 28.2% of 92

3. 60% of 68 is what number?

4. 25% of 96 is what number?

Estimate.

5. 8% of 40

6. 24% of 60

Write an equation for each problem. Then solve. Round to the nearest tenth if necessary.

7. What number is 8% of 50?

8. 52 is what percent of 260?

9. 30 is 75% of what number?

10. What is 15% of 24?

11. Tiffany answered 90% of the questions on her math test correctly. There were 50 questions on the test. How many questions did Tiffany answer correctly?

12. Jerilyn made 40 treats for her birthday. She gave 4 away to her family before taking the rest to school. What percent did she give away to her family?

1. _____
2. _____
3. _____
4. _____
5. _____
6. _____
7. _____
8. _____
9. _____
10. _____
11. _____
12. _____

NAME _____ DATE _____ PERIOD _____

Vocabulary Test

SCORE _____

discount	percent error	principal
gratuity	percent of change	sales tax
markdown	percent of decrease	selling price
markup	percent of increase	simple interest
percent equation	percent proportion	tip

Write the letter of the term that best matches each statement. You may use a term more than once.

_____ 1. the percent of change when the original quantity is greater than the new quantity

a. percent of increase

_____ 2. a ratio that compares a change in quantity to the original amount

b. percent of change

_____ 3. an equation (part = percent · base) in which the percent is written as a decimal

c. percent equation

_____ 4. the amount of money originally deposited, invested, or borrowed

d. sales tax

_____ 5. an amount of money charged by a government on items that people buy

e. simple interest

_____ 6. the amount by which the regular price of an item is reduced

f. percent of decrease

_____ 7. given by the formula $I = prt$

g. principal

_____ 8. the amount of money paid or earned on an investment or deposit for the use of the money

h. discount

_____ 9. the percent of change when the original quantity is less than the new quantity

i. tip

_____ 10. a small amount by which the regular price of a service is increased to express appreciation for the service

Define each term in your own words.

11. percent proportion

11. _____

12. gratuity

12. _____

Course 2 • Chapter 2 Percents

Standardized Test Practice

Read each question. Then fill in the correct answer on the answer sheet provided by your teacher or on a sheet of paper.

1. Sarah wants to buy new pillows for her room. Which store offers the best buy on pillows?

Store	Sale Price
A	3 pillows for $40
B	4 pillows for $50
C	2 pillows for $19
D	1 pillow for $11

   A. Store A
   B. Store B
   C. Store C
   D. Store D

2. The graph shows the attendance at a summer art festival from 2008 to 2013. If the trend in attendance continues, which of the following is a reasonable prediction for the attendance in 2017?

   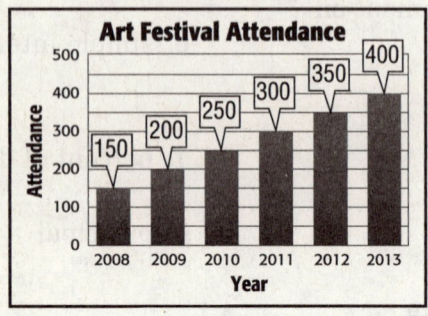

   F. Fewer than 200
   G. Between 500 and 600
   H. Between 700 and 800
   I. More than 800

3. At their annual car wash, the science club washes 30 cars in 45 minutes. At this rate, how many cars will they wash in 1 hour?

   A. 40
   B. 45
   C. 50
   D. 60

4. **GRIDDED RESPONSE** A necklace regularly sells for $18.00. The store advertises a 15% discount. What is the sale price of the necklace in dollars?

5. **GRIDDED RESPONSE** At a middle school, 38% of all seventh graders have taken swimming lessons. There are 250 students in the seventh grade. How many of them have taken swimming lessons?

6. The cost of Ken's car wash was $23.95. If he wants to give his detailer a 15% tip, about how much of a tip should he leave?

   F. $2.40
   G. $3.60
   H. $4.60
   I. $4.80

7. Cody has $700 in a savings account that pays 4% simple interest. What is the amount of simple interest he will earn in 2 years?

   A. $5.60
   B. $56
   C. $560
   D. $756

8. At a pet store, 38% of the animals are dogs. If there are a total of 88 animals at the pet store, which proportion can be used to find x, the number of dogs at the pet store?

   F. $\frac{x}{88} = \frac{100}{38}$

   G. $\frac{38}{88} = \frac{100}{x}$

   H. $\frac{x}{88} = \frac{38}{100}$

   I. $\frac{100}{88} = \frac{x}{38}$

9. **GRIDDED RESPONSE** A wrestler competes in 25 matches. Of those matches, he wins 17. What percent of the matches did the wrestler win?

10. **SHORT RESPONSE** The average cost per month of a 2-bedroom apartment in Grayson was $625 last year. This year, the average cost is $650. What is the percent of increase from last year to this year?

11. Mr. Cooper asked his students whether they prefer to go to the aquarium or the planetarium for a field trip. The table shows the results.

Response	Percent
Aquarium	50
Planetarium	25

    Suppose the rest of the class had no preference. What is the ratio of students who have no preference to the students who prefer to go to the aquarium?

    A. 1:5   C. 1:3
    B. 1:4   D. 1:2

12. In Nadia's DVD collection, she has 8 action DVDs, 12 comedy DVDs, 7 romance DVDs, and 3 science fiction DVDs. What percent of Nadia's DVD collection is comedies?

    F. 25%
    G. 30%
    H. 35%
    I. 40%

13. A salesman needed to sell a four-wheeler. He priced it at $3,500 the first day it was on the market. The second day he reduced the price by 10%. What was the price of the four-wheeler after this reduction?

    A. $3,850
    B. $3,465
    C. $3,150
    D. $3,000

14. **EXTENDED RESPONSE** Cable Company A increases their rates from $98 a month to $101.92 a month.

    Part A What is the percent of increase?

    Part B Cable Company B offers their cable for $110 dollars a month but gives a 10% discount for new customers. Describe two ways to find the cost for new customers.

    Part C If you currently use Cable Company A, would it make sense to change to Cable Company B? Explain.

Course 2 • Chapter 2 Percents

NAME _____ DATE _____ PERIOD _____

Student Recording Sheet

SCORE _____

Use this recording sheet with the Standardized Test Practice pages.

Fill in the correct answer. For gridded-response questions, write your answers in the boxes on the answer grid and fill in the bubbles to match your answers.

1. Ⓐ Ⓑ Ⓒ Ⓓ
2. Ⓕ Ⓖ Ⓗ Ⓘ
3. Ⓐ Ⓑ Ⓒ Ⓓ

4.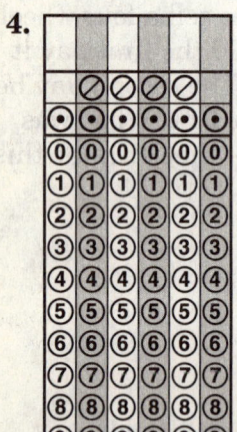

5.

6. Ⓕ Ⓖ Ⓗ Ⓘ
7. Ⓐ Ⓑ Ⓒ Ⓓ
8. Ⓕ Ⓖ Ⓗ Ⓘ

9.

10. _____
11. Ⓐ Ⓑ Ⓒ Ⓓ
12. Ⓕ Ⓖ Ⓗ Ⓘ
13. Ⓐ Ⓑ Ⓒ Ⓓ

Extended Response
Record your answers for Exercise 14 on the back of this paper.

34   Course 2 • Chapter 2 Percents

NAME _____ DATE _____ PERIOD _____

SCORE _____

Extended-Response Test

Demonstrate your knowledge by giving, a clear, concise solution to each problem. Be sure to include all relevant drawings and justify your answers. You may show your solutions in more than one way or investigate beyond the requirements of the problem. If necessary, record your answer on another piece of paper.

Most teens start their school shopping long before classes start, as shown on the graph.

1. Write an equation to find the number of teens out of 2,000 who said they shop in July. Solve. Show your work.

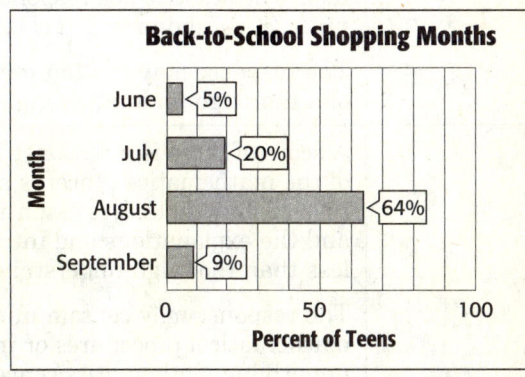

2. a. Find the rate of discount for a pair of pants that cost $65 and are on sale for $41.99. Explain each step.

   b. An item originally costs $16.95 and is on sale for 22% off the original price. The sales tax is 6.5%. Find the cost of the item. Explain each step.

3. a. Find the simple interest on $1,200 at 8% for 6 months. Explain each step.

   b. Find the interest on $900 at 6% for 4 years if the interest is added to the principal at the end of each year. Show your work.

Course 2 • Chapter 2 Percents

35

NAME _____ DATE _____ PERIOD _____

Extended-Response Rubric

SCORE _____

Score	Description
4	A score of four is a response in which the student demonstrates a thorough understanding of the mathematics concepts and/or procedures embodied in the task. The student has responded correctly to the task, used mathematically sound procedures, and provided clear and complete explanations and interpretations. The response may contain minor flaws that do not detract from the demonstration of a thorough understanding.
3	A score of three is a response in which the student demonstrates an understanding of the mathematics concepts and/or procedures embodied in the task. The student's response to the task is essentially correct with the mathematical procedures used and the explanations and interpretations provided demonstrating an essential but less than thorough understanding. The response may contain minor flaws that reflect inattentive execution of mathematical procedures or indications of some misunderstanding of the underlying mathematics concepts and/or procedures.
2	A score of two indicates that the student has demonstrated only a partial understanding of the mathematics concepts and/or procedures embodied in the task. Although the student may have used the correct approach to obtaining a solution or may have provided a correct solution, the student's work lacks an essential understanding of the underlying mathematical concepts. The response contains errors related to misunderstanding important aspects of the task, misuse of mathematical procedures, or faulty interpretations of results.
1	A score of one indicates that the student has demonstrated a very limited understanding of the mathematics concepts and/or procedures embodied in the task. The student's response is incomplete and exhibits many flaws. Although the student's response has addressed some of the conditions of the task, the student reached an inadequate conclusion and/or provided reasoning that was faulty or incomplete. The response exhibits many flaws or may be incomplete.
0	A score of zero indicates that the student has provided no response at all, or a completely incorrect or uninterpretable response, or demonstrated insufficient understanding of the mathematics concepts and/or procedures embodied in the task. For example, a student may provide some work that is mathematically correct, but the work does not demonstrate even a rudimentary understanding of the primary focus of the task.

NAME _____ DATE _____ PERIOD _____

SCORE _____

Test, Form 1A

Write the letter for the correct answer in the blank at the right of each question.

1. What is 138% of 250?
   A. 0.152   B. 345   C. 690   D. 950

   1. _____

2. What percent of 80 is 8?
   F. 0.1%   G. 1%   H. 10%   I. 72%

   2. _____

3. Twenty-four percent of the 25 swim team members are new on the team. How many members are new?
   A. 6   B. 8   C. 10   D. 12

   3. _____

Which is the best estimate for each expression?

4. 49% of 15
   F. 1.5   G. 4   H. 7.5   I. 9

   4. _____

5. $\frac{3}{4}$% of 387
   A. 300   B. 40   C. 3   D. 0.4

   5. _____

Which equation and solution represents each situation? Solutions may be rounded to the nearest tenth.

6. What number is 74% of 58?
   F. $0.74 = n \cdot 58$; 1.3
   G. $n = 0.74 \cdot 58$; 42.9
   H. $58 = n \cdot 0.74$; 78.4
   I. $n = 74 \cdot 58$; 4,292

   6. _____

7. 89% of what number is 14?
   A. $0.89 = n \cdot 14$; 0.1
   B. $14 = 89 \cdot n$; 0.2
   C. $n = 0.89 \cdot 14$; 12.5
   D. $14 = 0.89 \cdot n$; 15.7

   7. _____

8. While shopping, Lucinda spent $48. If the amount she spent was 15% of her savings, how much savings did she have before she shopped?
   F. $40.80   G. $55.20   H. $150   I. $320

   8. _____

9. Yul has lunch at a restaurant. If his bill was $14.50 and Yul wants to leave a 20% tip, what is the amount he should leave for the tip?
   A. $2.90   B. $2.50   C. $2.00   D. $1.44

   9. _____

10. An aquarium is on sale for $59.50. If this price represents a 15% discount from the original price, what is the original price to the nearest cent?
    F. $50.25   G. $60   H. $65.75   I. $70

    10. _____

Course 2 • Chapter 2 Percents

NAME _____ DATE _____ PERIOD _____

Test, Form 1A (continued)

SCORE _____

What is each percent of change? Round to the nearest whole percent if necessary. Is the percent of change an *increase* or *decrease*?

11. 100 to 150
    - A. 50%; decrease
    - B. 50%; increase
    - C. 33%; increase
    - D. $\frac{1}{2}$%; increase

    11. _____

12. $300 to $200
    - F. 100%; decrease
    - G. 50%; decrease
    - H. 33%; decrease
    - I. 33%; increase

    12. _____

13. 30 to 90
    - A. 3%; increase
    - B. 67%; increase
    - C. 200%; increase
    - D. 300%; increase

    13. _____

What is the total cost or sale price to the nearest cent?

14. $20 haircut; 15% tip
    - F. $3
    - G. $17
    - H. $23
    - I. $26

    14. _____

15. $10 lamp; 5% tax
    - A. $0.50
    - B. $9.50
    - C. $10.50
    - D. $15.00

    15. _____

16. $50 tool set; 10% markup
    - F. $5
    - G. $40
    - H. $45
    - I. $55

    16. _____

What is the simple interest paid to the nearest cent for each principal, interest rate, and time?

17. $1,000, 5%, 2 years
    - A. $10,000
    - B. $1,000
    - C. $100
    - D. $50

    17. _____

18. $300, $6\frac{1}{2}$%, 1 year
    - F. $19.50
    - G. $195
    - H. $319.50
    - I. $1,950

    18. _____

19. $850, 4%, 6 months
    - A. $2,040
    - B. $1,700
    - C. $204
    - D. $17

    19. _____

20. Chaleah deposited $900 in a new account that earns 6% simple interest. After 2 years, how much interest will she have earned?
    - F. $54
    - G. $108
    - H. $540
    - I. $1,080

    20. _____

NAME _____ DATE _____ PERIOD _____

Test, Form 1B

SCORE _____

Write the letter for the correct answer in the blank at the right of each question.

1. What is 164% of 25?
   A. 16   B. 28   C. 41   D. 75

   1. _____

2. What percent of 700 is 385?
   F. 0.55%   G. 1.82%   H. 45%   I. 55%

   2. _____

3. Fifteen percent of the 20 players on the soccer team are new this year. How many players on the team are new this year?
   A. 3   B. 6   C. 10   D. 15

   3. _____

Which is the best estimate for each expression?

4. 37% of 293
   F. 75   G. 120   H. 125   I. 150

   4. _____

5. $\frac{4}{5}$% of 192
   A. 2   B. 8   C. 12   D. 19

   5. _____

Which equation and solution represents each situation? Solutions may be rounded to the nearest tenth.

6. What number is 16% of 44?
   F. $n = 16 \cdot 44$; 704   H. $n = 0.16 \cdot 44$; 7.0
   G. $44 = 0.16 \cdot n$; 275   I. $0.16 = n \cdot 44$; 0.4

   6. _____

7. 36% of what number is 27?
   A. $27 = 0.36 \cdot n$; 75   C. $0.36 = n \cdot 27$; 1.3
   B. $n = 0.36 \cdot 27$; 9.7   D. $27 = 36 \cdot n$; 0.8

   7. _____

8. While shopping, Santiago spent $57. If the amount he spent was 30% of his savings, how much savings did he have before he shopped?
   F. $190   G. $171   H. $87   I. $39.90

   8. _____

9. Maria has lunch at a restaurant. If her bill was $12.25 and Maria wants to leave a 20% tip, what is a reasonable estimate for the amount she should leave for the tip?
   A. $1.30   B. $2.00   C. $2.45   D. $2.80

   9. _____

10. A keyboard is on sale for $676. If this price represents a 20% discount from the original price, what is the original price to the nearest cent?
    F. $650   G. $700   H. $845   I. $965

    10. _____

Course 2 • Chapter 2 Percents

NAME _____ DATE _____ PERIOD _____

Test, Form 1B (continued)   SCORE _____

What is each percent of change? Round to the nearest whole percent if necessary. Is the percent of change an *increase* or *decrease*?

11. 200 to 300
    - A. 33%; increase
    - B. 50%; decrease
    - C. 50%; increase
    - D. 100%; increase

    11. _____

12. $99 to $74
    - F. 34%; decrease
    - G. 34%; increase
    - H. 25%; decrease
    - I. 25%; increase

    12. _____

13. $49 to $149
    - A. 2%; increase
    - B. 67%; increase
    - C. 100%; increase
    - D. 204%; increase

    13. _____

What is the total cost or sale price to the nearest cent?

14. $1,725 couch; 15% markup
    - F. $1,983.75
    - G. $1,638.75
    - H. $1,466.25
    - I. $258.75

    14. _____

15. $14.30 watch; $6\frac{3}{4}$% sales tax
    - A. $0.97
    - B. $13.33
    - C. $15.27
    - D. $23.95

    15. _____

16. $48 dinner; 20% tip
    - F. $9.60
    - G. $57.60
    - H. $62.60
    - I. $68.00

    16. _____

What is the simple interest paid to the nearest cent for each principal, interest rate, and time?

17. $2,500, 4.5%, 2 years
    - A. $225
    - B. $2,250
    - C. $2,725
    - D. $22,500

    17. _____

18. $834, 3%, 15 months
    - F. $31.28
    - G. $375.30
    - H. $865.28
    - I. $3,127.50

    18. _____

19. $1,750, $5\frac{3}{4}$%, 9 months
    - A. $70.09
    - B. $75.47
    - C. $905.63
    - D. $1,825.47

    19. _____

20. Roberto deposited $860 in a new account that earns 6.5% simple interest. After 6 months, how much interest will he have earned?
    - F. $279.50
    - G. $55.90
    - H. $27.95
    - I. $2.80

    20. _____

NAME _____ DATE _____ PERIOD _____

Test, Form 2A

SCORE _____

Write the letter for the correct answer in the blank at the right of each question.

1. What is 25% of 40?
   A. 5   B. 10   C. 20   D. 50

   1. _____

2. What is 115% of 80?
   F. 12   G. 15   H. 85   I. 92

   2. _____

3. 8% of 70 is what number?
   A. 5.6   B. 10.4   C. 15.1   D. 78

   3. _____

4. Angela made 90% of the 50 free throws she attempted. How many free throws did Angela make?
   F. 10   G. 27   H. 41   I. 45

   4. _____

Which is the best estimate for each of the following?

5. 26% of 44
   A. 5   B. 11   C. 18   D. 21

   5. _____

6. 0.5% of 600
   F. 1.4   G. 2.5   H. 3   I. 6

   6. _____

Which equation and solution represents each situation?

7. What number is 32% of 79?
   A. $n = 32 \cdot 79$; 2,528
   B. $32 = n \cdot 79$; 0.4
   C. $79 = n \cdot 0.32$; 246.9
   D. $n = 0.32 \cdot 79$; 25.3

   7. _____

8. 10% of what number is 30?
   F. $n = 0.1 \cdot 30$; 3
   G. $30 = n \cdot 10$; 3
   H. $30 = 0.1 \cdot n$; 300
   I. $n = 30 \cdot 1.0$; 30

   8. _____

9. What number is 64% of 120?
   A. $n = 64 \cdot 120$; 7,680
   B. $64 = n \cdot 120$; 0.5
   C. $n = 0.64 \cdot 120$; 76.8
   D. $n = 1.20 \cdot 0.64$; 0.8

   9. _____

Course 2 • Chapter 2 Percents

NAME _____ DATE _____ PERIOD _____

Test, Form 2A (continued)   SCORE _____

10. While eating lunch, Zeshon ate 75% of the carrots he had in his bag. There were 12 carrots in his bag. How many carrots did Zeshon eat?

    F. 6   G. 9   H. 10   I. 12   10. _____

11. Melinda ate breakfast and wanted to leave her waitress a 20% tip. Her meal cost $7.50. How much money should Melinda leave for the tip?   11. _____

Find each percent of change. Round to the nearest whole percent if necessary. State whether the percent of change is an *increase* or *decrease*.

12. Zach bought a pair of jeans for $54. The next week he noticed that the price for the same pair of jeans was now $74. Find the percent of change.   12. _____

13. 40 to 25   13. _____

14. 120 to 140   14. _____

Find the simple interest paid to the nearest cent for each principal, interest rate, and time.

15. $250, 3.4%, 3 years   15. _____

16. $570, 2%, 4 years   16. _____

17. Darin bought a new pair of soccer cleats for $59.99. The sales tax is 6.5%. What is the total cost that Darin will pay?   17. _____

18. Mr. Martin bought a microwave that was originally priced at $225. He received a 30% discount. What is the sale price of the microwave?   18. _____

19. Yancy got his hair cut for $16. He left a 15% tip for the barber. What is the total amount of money Yancy paid?   19. _____

42   Course 2 • Chapter 2 Percents

NAME _____ DATE _____ PERIOD _____

SCORE _____

Test, Form 2B

Write the letter for the correct answer in the blank at the right of each question.

1. What is 35% of 60?
   A. 7   **B.** 12   **C.** 21   **D.** 28

   1. _____

2. What is 112% of 50?
   F. 56   **G.** 52   **H.** 48   **I.** 6

   2. _____

3. 3% of 120 is what number?
   A. 3.6   **B.** 4.4   **C.** 9.5   **D.** 40

   3. _____

4. Kaitlyn made 85% of the 60 free throws she attempted. How many free throws did Kaitlyn make?
   F. 17   **G.** 25   **H.** 51   **I.** 60

   4. _____

Which is the best estimate for each of the following?

5. 23% of 36
   A. 3   **B.** 6   **C.** 9   **D.** 15

   5. _____

6. 0.8% of 503
   F. 10   **G.** 7.5   **H.** 4   **I.** 1

   6. _____

Which equation and solution represents each situation?

7. What number is 16% of 46?
   A. $n = 16 \cdot 46$; 736
   B. $16 = n \cdot 46$; 34.8
   C. $46 = n \cdot 0.16$; 287.5
   D. $n = 0.16 \cdot 46$; 7.4

   7. _____

8. 20% of what number is 30?
   F. $n = 0.2 \cdot 30$; 6
   G. $30 = 0.2 \cdot n$; 150
   H. $30 = n \cdot 20$; 1.5
   I. $n = 30 \cdot 2.0$; 60

   8. _____

9. What number is 76% of 140?
   A. $n = 76 \cdot 140$; 10,640
   B. $76 = n \cdot 140$; 0.5
   C. $n = 0.76 \cdot 140$; 106.4
   D. $n = 1.40 \cdot 0.76$; 1.1

   9. _____

Course 2 • Chapter 2 Percents

NAME _____ DATE _____ PERIOD _____

Test, Form 2B (continued)

SCORE _____

10. While eating dinner, Marian ate 25% of the apple slices she had in her bag. There were 20 apple slices in her bag. How many apple slices did Marian eat?

    F. 5   G. 7   H. 9   I. 10

10. _____

11. Adalia ate lunch and wanted to leave her waitress a 15% tip. Hermeal cost $8.95. How much money should Adalia leave for the tip?

11. _____

Find each percent of change. Round to the nearest whole percent if necessary. State whether the percent of change is an *increase* or *decrease*.

12. Wyatt bought a pair of shoes for $72. The next week he noticed that the price for the same pair of shoes was now $87. Find the percent of change.

12. _____

13. 55 to 35

13. _____

14. 170 to 210

14. _____

Find the simple interest paid to the nearest cent for each principal, interest rate, and time.

15. $1,450, 2.5%, 2 years

15. _____

16. $390, 3%, 5 years

16. _____

17. Eber bought a new pair of baseball cleats for $64.95. The sales tax is 5.5%. What is the total cost that Eber will pay?

17. _____

18. Mr. Jamison bought a toaster oven that was originally priced at $89. He received a 20% discount. What is the sale price of the toaster oven?

18. _____

19. Winona got her hair cut for $21. She left a 15% tip for the hair stylist. What is the total amount of money Winona paid?

19. _____

Course 2 • Chapter 2 Percents

NAME _____ DATE _____ PERIOD _____

Test, Form 3A

SCORE _____

Find each number.

1. 20% of 120 is what number?

   1. _____

2. 15 is 15% of what number?

   2. _____

3. 8 is what percent of 64?

   3. _____

4. Estimate 60% of 52.

   4. _____

5. Estimate $\frac{2}{5}$ of 128.

   5. _____

6. Estimate 215% of 18.

   6. _____

For Exercises 7 and 8, write an equation for each problem. Then solve. Round to the nearest tenth.

7. 15 is what percent of 48?

   7. _____

8. 35% of what number is 42?

   8. _____

9. Macy has completed 78% of the 50 questions on the test. How many questions has Macy completed?

   9. _____

10. Christina estimates that she will spend $250 on new summer clothes. She actually spent $350. What is the percent error? Round to the nearest whole percent.

    10. _____

11. Gregory predicts that 310 people will attend the spring play. There was an actual total of 220 people who attended the spring play. What is the percent error? Round to the nearest whole percent.

    11. _____

12. Emilio and Regina each sell popcorn at the basketball games. The table shows how much they charge for the popcorn including markup for supplies and the number of bags popcorn sold. Who made more profit? Explain.

	Cost of Popcorn	Markup	Number of Bags Sold
Emilio	$1.25	8%	30
Regina	$1.40	5%	26

    12. _____

Course 2 • Chapter 2 Percents

45

NAME _____ DATE _____ PERIOD _____

Test, Form 3A (continued)   SCORE _____

Find each percent of change. Round to the nearest whole percent. State whether the percent of change is an *increase* or a *decrease*.

13. $75 to $25   13. _____

14. 7,500 to 8,200   14. _____

15. Kenzie bought a shirt for $18. The next day she saw the shirt was selling for $24.60. What is the percent of change?   15. _____

16. Last year, Sandro bought a book for $13.89. This year the same book cost $15.79. What was the percent of change?   16. _____

17. The regular price of a new washer and dryer is $975 and the sale price is $850. Find the percent of decrease to the nearest whole percent.   17. _____

Find the total cost or sale price to the nearest cent.

18. $6.95 lunch; 20% tip   18. _____

19. $30 bat; 15% discount   19. _____

20. $2,500 jet ski; 6.5% tax   20. _____

21. $12.95 dinner; 15% tip   21. _____

Find the simple interest paid to the nearest cent for each principal, interest rate, and time.

22. $1,400, 3.8%, 1 year   22. _____

23. $628, 5%, 4 months   23. _____

24. $85, 4.5%, 3 years   24. _____

25. Mr. Fraser bought new shingles for his house for $3,500 using a credit card. His card has an interest rate of 17%. If he does not pay off his balance at the end of the month and has no other charges, how much money will he owe on his credit card at the end of the month?   25. _____

Course 2 • Chapter 2 Percents

Test, Form 3B

Find each number.

1. 16% of 95 is what number?

   1. _____

2. 47 is 0.5% of what number?

   2. _____

3. 6 is what percent of 96?

   3. _____

4. Estimate $\frac{1}{5}$% of 201.

   4. _____

5. Estimate 19% of 112.

   5. _____

6. Estimate 306% of 25.

   6. _____

For Exercises 7 and 8, write an equation for each problem. Then solve. Round to the nearest tenth.

7. 14 is what percent of 70?

   7. _____

8. 65% of what number is 39?

   8. _____

9. Kelsey read 75% of the 40 books she bought. How many books has Kelsey completed?

   9. _____

10. Victor estimates that he will spend $175 on new video games. He actually spent $225. What is the percent error? Round to the nearest whole percent.

    10. _____

11. Sheila predicts that 480 people will attend the fall concert. There was an actual total of 350 people who attended the fall concert. What is the percent error? Round to the nearest whole percent.

    11. _____

12. Suzanne and Michael each sell slices of pizza at the basketball games. The table shows how much they charge for a slice of pizza including markup for supplies and the number of slices sold. Who made more profit? Explain.

	Cost of Popcorn	Markup	Slices of Pizza Sold
Suzanne	$2.00	5%	34
Michael	$2.25	4%	30

    12. _____

Course 2 • Chapter 2 Percents

Test, Form 3B (continued)

Find each percent of change. Round to the nearest whole percent. State whether the percent of change is an *increase* or a *decrease*.

13. $210 to $35

14. 6,520 to 9,408

15. The price of a dozen cookies at a bake sale last year was $2. This year the price for a dozen of cookies was $5. What is the percent of change?

16. Five years ago, one share of a stock was worth $28.00. Today, one share of the stock is worth $62.00. What was the percent of change?

17. If the regular price of a new T-shirt is $13.95 and the sale price is $10.00, find the percent of decrease to the nearest whole percent.

Find the total cost or sale price to the nearest cent.

18. $14.95 dinner; 15% tip

19. $27.99 shoes; $7\frac{1}{2}$% tax

20. $16.99 cap; 20% discount

21. $25.00 game ticket; 5.5% tax

Find the simple interest paid to the nearest cent for each principal, interest rate, and time.

22. $2,620, 3%, 5 years

23. $152, 2.5%, 18 months

24. $72.80, $4\frac{1}{4}$%, 2 years

25. Mrs. Martin deposits $2,350 in a money market account. Her account earns an interest rate of 18%. If she does not deposit or withdraw any money from the account, how much will Mrs. Martin have in the account at the end of six months?

NAME _____ DATE _____ PERIOD _____

Are You Ready?

Review

Order of Operations
1. Simplify the expressions inside grouping symbols, like parentheses.
2. Find the value of all powers.
3. Multiply and divide in order from left to right.
4. Add and subtract in order from left to right.

Example 1
Evaluate $56 \div 7 + 3 \cdot 4$.

$$\begin{aligned} 56 \div 7 + 3 \cdot 4 &= 8 + 3 \cdot 4 &\text{Divide 56 by 7.} \\ &= 8 + 12 &\text{Multiply 3 by 4.} \\ &= 20 &\text{Add.} \end{aligned}$$

Example 2
Evaluate $2^3 \cdot 5 + 4$.

$$\begin{aligned} 2^3 \cdot 5 + 4 &= 8 \cdot 5 + 4 &\text{Find } 2^3. \\ &= 40 + 4 &\text{Multiply 8 by 5.} \\ &= 44 &\text{Add.} \end{aligned}$$

Exercises

Evaluate.

1. $64 \div (9 + 7)$

2. $37 + 42 - 28$

3. $6^2 + 40 \div 8$

4. $32 + 16 - 40$

5. $16 + 3(5 - 2)$

6. $(6 + 8 \cdot 2) \div 11$

7. $(42 \div 7) + (4 - 1)$

8. $9 + (54 \div 9)$

9. $7 + 4 + (11 + 2)$

10. $12 + 4 - 2$

Course 2 • Chapter 3 Integers

Are You Ready?

Practice

Evaluate.

1. $48 \div (10 + 6)$
2. $38 + 51 - 29$
3. $15 + 14 - 19$
4. $30 + 3(6 - 2)$
5. $(5 + 7 \times 3) \div 13$
6. $(24 \div 6) + (5 - 3)$
7. $8 + 18 \div 9$
8. $11 + (36 \div 4)$

Use the coordinate plane to name the ordered pair for each point.

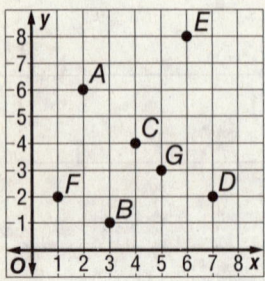

9. A
10. B
11. C
12. D
13. E
14. F
15. G

NAME _____ DATE _____ PERIOD _____

Are You Ready?

Apply

1. **BASEBALL** John had 30 baseball cards. He gave 14 cards to Mike and 7 to Jeff. How many baseball cards does John have left?

2. **PARTIES** Louise is having a party. She bought 96 pieces of chicken. If she planned to have 3 pieces for each guest, how many people could she invite?

3. **FUNDRAISER** The soccer team collected soda cans for a fundraiser. They had 175 cans and found 58 more. The next day, they turned in 97 cans. How many cans do they have left?

4. **CORN** Mr. Rodriguez planted 22 rows of corn. There were 15 plants in each row. How many corn plants did he put in his garden?

5. **MAPS** The graph below shows the locations of four places. What are the coordinates of the skating rink?

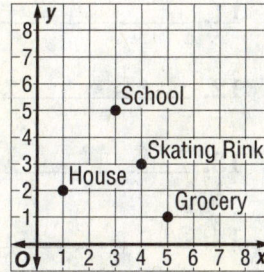

6. **MAPS** Refer to Exercise 5. The library is located at (2, 4). Which location on the map is closest to the library?

Course 2 • Chapter 3 Integers   51

Diagnostic Test

Evaluate.

1. $37 + 52 - 46$
2. $(5 + 11 \times 4) \div 7$
3. $36 \div (3 + 1)$
4. $8 + (54 \div 9)$
5. $6 + 40 \div 8$
6. $(30 \div 5) \div (4 - 2)$
7. $50 + 10 - 41$
8. $12 + 7(5 - 2)$

1. _____
2. _____
3. _____
4. _____
5. _____
6. _____
7. _____
8. _____

Use the coordinate plane to name the ordered pair for each point.

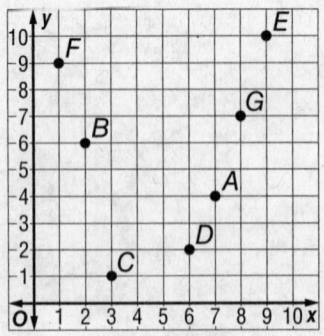

9. A
10. B
11. C
12. D
13. E
14. F
15. G

9. _____
10. _____
11. _____
12. _____
13. _____
14. _____
15. _____

Pretest

1. Graph the set of integers {−2, 1, −4} on a number line.

    1. ←+++++++++++→

2. Graph the set of integers {7, −1, 3, −2} on a number line.

    2. ←+++++++++++++→

Evaluate each expression.

3. | 11 |

    3. _____

4. 4 + | −6 |

    4. _____

5. | −7 | − | −4 |

    5. _____

6. $a - b$ if $a = 4$ and $b = 12$

    6. _____

Add, subtract, multiply, or divide.

7. 8 + (−4)

    7. _____

8. −7 + 2

    8. _____

9. −11 − 7

    9. _____

10. 12 ÷ (−6)

    10. _____

11. −9 × (−3)

    11. _____

12. 4 − (−3)

    12. _____

13. −45 ÷ 9

    13. _____

14. 6 × (−3)

    14. _____

15. **SAVINGS ACCOUNT** Ms. Cole pays $2 a month for online banking at her credit union. What is the change in the balance of her account if she pays this fee for 6 months?

    15. _____

Course 2 • Chapter 3 Integers

Chapter Quiz

1. Graph the set of integers {−2, −5, 1} on a number line.

2. Graph the set of integers {4, −3, 2} on a number line.

Write an integer for each situation.

3. 15°F below 0

4. a deposit of $24

Evaluate each expression.

5. $|-3|$

6. $|9|$

7. $|13| + |-2|$

8. $|-12| - |8|$

Add or subtract.

9. $12 + (-7)$

10. $-9 + 8$

Evaluate each expression if $f = -4$, $g = 2$, and $h = 7$.

11. $-h - 3$

12. $h - f$

13. **TUNNEL** A mine worker is in a tunnel 25 feet below the ground. He descends another 13 feet. What is his final position?

1. _____
2. _____
3. _____
4. _____
5. _____
6. _____
7. _____
8. _____
9. _____
10. _____
11. _____
12. _____
13. _____

NAME _____ DATE _____ PERIOD _____

Vocabulary Test

absolute value	integer	opposites
additive inverse	negative integer	positive integer
graph		

Choose from the terms above to complete each sentence.

1. The distance a number is from zero on a number line is the _____ of the number.

   1. _____

2. A(n) _____ is any number from the set {…,−3, −2, −1, 0, 1, 2, 3, …}.

   2. _____

3. The sum of any number and its _____ equals 0.

   3. _____

4. A(n) _____ is an integer less than zero.

   4. _____

5. Two integers that are the same distance from 0 on a number line, but on opposite sides of 0, are called _____.

   5. _____

6. When you _____ a point on a number line, you locate its position by drawing a dot at the location of its value.

   6. _____

7. A(n) _____ is an integer greater than zero.

   7. _____

Course 2 • Chapter 3 Integers

Standardized Test Practice

Read each question. Then fill in the correct answer on the answer document provided by your teacher or on a sheet of paper.

1. The table shows the daily low temperatures for Cleveland, Ohio, over five days.

Day	Temperature
1	15°F
2	−2°F
3	8°F
4	−6°F
5	5°F

   Which expression can be used to find the average daily low temperature during the five days?

   A. $(15 + 2 + 8 + 6 + 5) \div 5$
   B. $15 + 2 + 8 + 6 + 5 \div 5$
   C. $[15 + (-2) + 8 + (-6) + 5] \div 5$
   D. $15 + (-2) + 8 + (-6) + 5 \div 5$

2. Three vertices of a parallelogram are given as coordinates (−4, 2), (−2, 4), and (1, −3) in the graph. Which coordinates best represent the location of the fourth vertex of the parallelogram?

   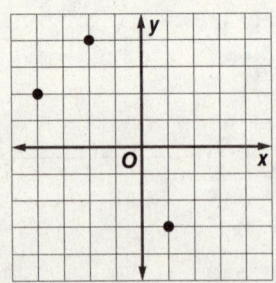

   F. (−3, 1)
   G. (3, −1)
   H. (1, −3)
   I. (−1, 3)

3. **GRIDDED RESPONSE** The lowest point in Japan is Hachiro-gata (elevation −4 m), and the highest point is Mount Fuji (elevation 3,776 m). What is the difference in elevation, in meters, between Mount Fuji and Hachiro-gata?

4. **GRIDDED RESPONSE** A submarine is cruising 8 meters below the surface. The captain orders a dive of another 17 meters. What is the new cruising depth of the submarine in meters?

5. In what quadrant is point P located?

   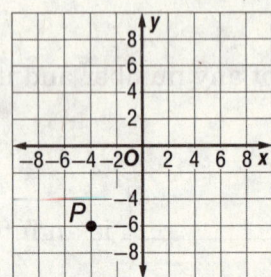

   A. Quadrant I   **C.** Quadrant III
   B. Quadrant II **D.** Quadrant IV

6. What integer added to −9 gives a sum of 3?

   F. 12   **H.** 3
   G. 6   **I.** −12

7. By the end of the third quarter of a football game, Ricky had gained 112 yards and had lost 12 yards. If Ricky lost an additional 8 yards and gained 22 yards in the fourth quarter, which equation could be used to represent his total yardage for the game?

   A. $112 + 12 + 8 + 22 = 154$
   B. $112 + (-12) + (-8) + 22 = 114$
   C. $112 + 12 + (-8) + (-22) = 94$
   D. $(-112) + (-12) + 8 + 22 = -94$

8. **GRIDDED RESPONSE** Bobby is diving 50 feet below sea level at the beach. His sister is at the swimming pool deck, which is 15 feet above sea level. What is the difference, in feet, between the pool deck and Bobby's position?

9. **SHORT RESPONSE** Larry borrowed $12,000 from his grandfather to buy a car. He bought a used car, so he returned $4,411 to his grandfather. Write and solve an equation using integers that shows the total amount that Larry owes his grandfather.

10. Pablo and three of his friends are playing paintball. The table shows their scores at the end of one round. By how many points is Winston beating Pablo?

Player	Score
Pablo	−189
Winston	−124
Nevin	130
Marsella	48

    F. 65
    G. 135
    H. 178
    I. 313

11. Each of the first 4 pit stops a race car driver makes loses ten seconds off the leader. The pit crew makes adjustments, and at each of the next two pit stops he gains 7 seconds on the leader. How much time is the driver off the leader?

    A. 40 seconds
    B. 14 seconds
    C. −26 seconds
    D. −54 seconds

12. **SHORT RESPONSE** A rectangle and a square are graphed on a coordinate plane. Name an ordered pair that is inside the rectangle but outside the square.

    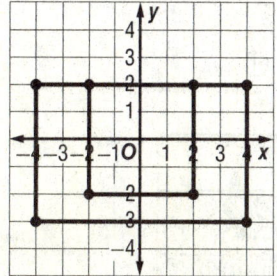

13. **EXTENDED RESPONSE** The use of computers to download music has decreased the sales of music CDs. Use the following table to answer the questions.

Year	Estimated Number of New Music CD Releases
2000	36,000
2001	32,000
2002	34,000
2003	8,000
2004	14,000
2005	10,000
2006	12,000

    Part A During which year was there the greatest decrease in CD releases from the previous year? What was the decrease?

    Part B Write and evaluate an expression that shows the change in CD releases from 2004 to 2005.

Course 2 • Chapter 3 Integers

NAME _____ DATE _____ PERIOD _____

Student Recording Sheet

SCORE _____

Use this recording sheet with the Standardized Test Practice pages.

Fill in the correct answer. For gridded-response questions, write your answers in the boxes on the answer grid and fill in the bubbles to match your answers.

1. Ⓐ Ⓑ Ⓒ Ⓓ

2. Ⓕ Ⓖ Ⓗ Ⓘ

3. [grid]

4. [grid]

5. Ⓐ Ⓑ Ⓒ Ⓓ

6. Ⓕ Ⓖ Ⓗ Ⓘ

7. Ⓐ Ⓑ Ⓒ Ⓓ

8. [grid]

9. _____

10. Ⓕ Ⓖ Ⓗ Ⓘ

11. Ⓐ Ⓑ Ⓒ Ⓓ

12. _____

Extended Response

Record your answers for Exercise 13 on the back of this paper.

58 Course 2 • Chapter 3 Integers

NAME _____ DATE _____ PERIOD _____

Extended-Response Test

SCORE _____

Demonstrate your knowledge by giving a clear, concise solution to each problem. Be sure to include all relevant drawings and justify your answers. You may show your solutions in more than one way or investigate beyond the requirements of the problem. If necessary, record your answer on another piece of paper.

1. Horatio, Glen, Carlos, and Meredith played in the company golf tournament. Their scores after the first round are listed below.

    Horatio: two over par (+2)
    Glen: three under par (−3)
    Carlos: four over par (+4)
    Meredith: one under par (−1)

    a. Explain what is meant by the *absolute value* of a number. Then find the absolute value of Glen's score.

    b. Glen's second round score is +2. Use counters to find Glen's total score after two rounds.

    c. Meredith's second round score is −3. Find her total score after two rounds.

2. The first year, an SUV was worth $43,500. Five years later, it was worth $13,000.

    a. If C is the current value and S is the starting value, write an expression to find the average change in value each year.

    b. Find the average change in value each year. Show your work.

3. Describe how opposites can be used when subtracting integers. Give an example to support your answer.

Course 2 • Chapter 3 Integers

NAME _____ DATE _____ PERIOD _____

Extended-Response Rubric

SCORE _____

Score	Description
4	A score of four is a response in which the student demonstrates a thorough understanding of the mathematics concepts and/or procedures embodied in the task. The student has responded correctly to the task, used mathematically sound procedures, and provided clear and complete explanations and interpretations. The response may contain minor flaws that do not detract from the demonstration of a thorough understanding.
3	A score of three is a response in which the student demonstrates an understanding of the mathematics concepts and/or procedures embodied in the task. The student's response to the task is essentially correct with the mathematical procedures used and the explanations and interpretations provided demonstrating an essential but less than thorough understanding. The response may contain minor flaws that reflect inattentive execution of mathematical procedures or indications of some misunderstanding of the underlying mathematics concepts and/or procedures.
2	A score of two indicates that the student has demonstrated only a partial understanding of the mathematics concepts and/or procedures embodied in the task. Although the student may have used the correct approach to obtaining a solution or may have provided a correct solution, the student's work lacks an essential understanding of the underlying mathematical concepts. The response contains errors related to misunderstanding important aspects of the task, misuse of mathematical procedures, or faulty interpretations of results.
1	A score of one indicates that the student has demonstrated a very limited understanding of the mathematics concepts and/or procedures embodied in the task. The student's response is incomplete and exhibits many flaws. Although the student's response has addressed some of the conditions of the task, the student reached an inadequate conclusion and/or provided reasoning that was faulty or incomplete. The response exhibits many flaws or may be incomplete.
0	A score of zero indicates that the student has provided no response at all, or a completely incorrect or uninterpretable response, or demonstrated insufficient understanding of the mathematics concepts and/or procedures embodied in the task. For example, a student may provide some work that is mathematically correct, but the work does not demonstrate even a rudimentary understanding of the primary focus of the task.

Test, Form 1A

Write the letter for the correct answer in the blank at the right of each question.

1. Which integer represents 8°C below 0?
   A. −8   B. 8   C. |−8| D. |8|

2. What is the value of |3|?
   F. −3   G. 0   H. 3   I. 6

3. What is the value of |−9|?
   A. 18   B. 9   C. 0   D. −9

4. Which integers represent S and T on the number line?
   F. S, 4; T, −2   H. S, −4; T, 2
   G. S, 2; T, −4   I. S, −2; T, 4

   ```
          S         T
   ←——+—+—+—●—+—+—+—+—●—+—→
     −6 −5 −4 −3 −2 −1 0 1 2 3 4 5
   ```

5. What is the value of 3 − 14?
   A. 17   B. 11   C. −11   D. −17

6. A turtle dives towards deeper water at a rate of 8 inches per second. It continues for a total of 16 seconds. Which expression represents this situation?
   F. 16(−8)   G. −16(−8)   H. 16 ÷ 8   I. 16 ÷ (−8)

7. Marisa has $26 in her purse. She pays $5 for lunch. Which expression represents this situation?
   A. −26 + (−5)   B. −26 + 5   C. 26 + (−5)   D. 26 + 5

8. Aaron turned on his air conditioning and the temperature in his apartment decreased 6 degrees. Write an integer to represent the change in temperature.
   F. 6°   G. −6°   H. 3°   I. −3°

9. Luisa is 43 feet underground touring a cavern. She climbs a ladder up 14 feet. What is her new location?
   A. −43 feet   B. −14 feet   C. −29 feet   D. −57 feet

10. The highest point in California is Mount Whitney at 14,494 feet above sea level. The lowest point in California is in Death Valley at 282 feet below sea level. What is the difference in elevations?
    F. 14,212 feet   G. 14,494 feet   H. 14,776 feet   I. 15,226 feet

Course 2 • Chapter 3 Integers

Test, Form 1A (continued)

11. An archaeologist descends 20 feet into a canyon, then climbs up 12 feet. Which value represents her final position?
    A. −32 ft   B. −12 ft   C. −8 ft   D. 8 ft

    11. _____

What is the value of each expression?

12. $8 + (-7)$
    F. 15   G. 1   H. −1   I. −15

    12. _____

13. $-7(-6)$
    A. 42   B. −1   C. −13   D. −42

    13. _____

14. $18 \div (-9)$
    F. 9   G. 2   H. −2   I. −9

    14. _____

15. $35 - 12$
    A. 47   B. 23   C. −23   D. −47

    15. _____

16. $(-3)^2$
    F. −9   G. −6   H. −1   I. 9

    16. _____

What is the value of each expression if $a = -4$, $b = 6$, and $c = -1$?

17. bc
    A. 6   B. −6   C. 5   D. −5

    17. _____

18. $10 - a$
    F. −6   G. 6   H. 14   I. −14

    18. _____

19. $9 + b$
    A. 54   B. −3   C. 3   D. 15

    19. _____

20. $\dfrac{-12}{b}$
    F. −18   G. −6   H. −2   I. 2

    20. _____

Test, Form 1B

Write the letter for the correct answer in the blank at the right of each question.

1. Which integer represents 15°C below 0?
   A. $|15|$ B. $|-15|$ C. 15   D. -15

   1. _____

2. What is the value of $|-6|$?
   F. -6 G. 0   H. $\frac{1}{6}$ I. 6

   2. _____

3. What is the value of $|5|$?
   A. 5   B. $\frac{1}{5}$ C. 0   D. -5

   3. _____

4. Which integers represent A and B on the number line?

   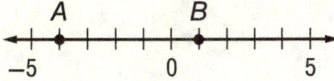

   F. $A, -1; B, 4$ G. $A, 4; B, -1$ H. $A, 1; B, -4$ I. $A, -4; B, 1$

   4. _____

5. What is the value of $5 - 17$?
   A. 22   B. 12   C. -12 D. -22

   5. _____

6. A submarine dives 10 feet per minute for 12 minutes. Which expression represents this situation?
   F. $12(10)$ G. $12(-10)$ H. $12 \div 10$ I. $12 \div (-10)$

   6. _____

7. Sandra has $32 in her purse. She pays $15 for a CD. Which expression represents this situation?
   A. $32 + 15$ B. $-32 + 15$ C. $-32 + (-15)$ D. $32 + (-15)$

   7. _____

8. Patrick turned on his air conditioning and the temperature in his apartment decreased 4 degrees. Write an integer to represent the change in temperature.
   F. $-4°$ G. $-2°$ H. $2°$ I. $4°$

   8. _____

9. Chi is 38 feet underground touring a cavern. He goes down a ladder 7 feet. What is his new location?
   A. -45 feet   B. -38 feet   C. -31 feet   D. -7 feet

   9. _____

10. The highest point in a state park is 145 feet above sea level. The lowest point is 28 feet below sea level. What is the difference in elevations?
    F. 117 feet   G. 28 feet   H. 145 feet   I. 173 feet

    10. _____

Course 2 • Chapter 3 Integers

NAME _____ DATE _____ PERIOD _____

Test, Form 1B (continued)

SCORE _____

11. An archaeologist descends 30 feet into a tomb, then climbs up 18 feet. Which value represents his final position?

    A. −48 ft   B. −12 ft   C. 12 ft   D. 48 ft

11. _____

What is the value of each expression?

12. $10 + (-8)$

    F. −18   G. −2   H. 2   I. 18

12. _____

13. $-8(-4)$

    A. 32   B. 12   C. −12   D. −32

13. _____

14. $24 \div (-6)$

    F. 18   G. 4   H. −4   I. −30

14. _____

15. $19 - 11$

    A. 30   B. 8   C. −8   D. −30

15. _____

16. $(-4)2$

    F. 16   G. 8   H. −8   I. −16

16. _____

What is the value of each expression if $m = -3$, $n = 8$, and $p = -1$?

17. np

    A. −9   B. −8   C. 7   D. 8

17. _____

18. $9 - m$

    F. −27   G. −6   H. 6   I. 12

18. _____

19. $-12 + n$

    A. 20   B. 4   C. −4   D. −20

19. _____

20. $\dfrac{-18}{m}$

    F. 6   G. −6   H. −15   I. −21

20. _____

Test, Form 2A

Write the letter for the correct answer in the blank at the right of each question.

1. Which integer represents a 5-yard loss?
   A. $|5|$ B. $|-5|$ C. 5   D. -5

   1. _____

2. What is the value of $|-7|$?
   F. 7   G. $-|-7|$ H. 14   I. -7

   2. _____

3. What is the value of $|5| + |-2|$?
   A. -7 B. -3 C. 3   D. 7

   3. _____

4. Which integers represent A and B on the number line?
   F. $A, -3; B, -5$ H. $A, 3; B, -5$
   G. $A, -3; B, 5$ I. $A, -5; B, 3$

   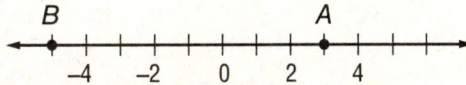

   4. _____

5. Which value of q makes $-15 - (-9) = q$ a true sentence?
   A. -24 B. -6 C. 6   D. 24

   5. _____

6. A snake dives towards deeper water at a rate of 14 inches per second. If the snake continues at this rate for a total of 6 seconds, which expression represents this situation?
   F. $6(-14)$ G. $-6(-14)$ H. $6 \div 14$ I. $6 \div (-14)$

   6. _____

7. Ishi has $24 in his wallet. He pays $8 to have his car washed. Which expression represents this situation?
   A. $-24 + (-8)$ B. $-24 + 8$ C. $24 + (-8)$ D. $24 + 8$

   7. _____

8. The highest point in Louisiana is Driskill Mountain at 585 feet. The lowest point is -8 feet in New Orleans. What is the difference in feet between the highest and lowest elevation points?
   F. 578 feet   G. 593 feet   H. 601 feet   I. 609 feet

   8. _____

9. The average daytime temperature on Venus is 870°F. The average temperature on Jupiter is -160°F. What is the difference between the average temperatures on Venus and Jupiter?
   A. 710°F   B. 730°F   C. 1,030°F   D. 1,050°F

   9. _____

Course 2 • Chapter 3 Integers   65

NAME _____ DATE _____ PERIOD _____

Test, Form 2A (continued)   SCORE _____

10. An archaeologist descends 22 feet into a canyon, then climbs up 15 feet. What is his final position?

F. −37 ft   G. −15 ft   H. −7 ft   I. 7 ft   10. _____

What is the value of each expression?

11. $12 + (-5)$
A. −17   B. −7   C. 7   D. 17   11. _____

12. $-8(-10)$
F. −80   G. −18   H. 18   I. 80   12. _____

13. $\frac{-48}{-24}$
A. 2   B. 0.5   C. −2   D. −24   13. _____

14. $-37 - 8$
F. −45   G. −29   H. 45   I. 296   14. _____

15. $7(-3)$
A. 21   B. 4   C. $-2\frac{1}{3}$ D. −21   15. _____

Evaluate each expression if $x = -3$, $y = 8$, and $z = -4$.

16. $y - (-5)$ 16. _____

17. xz 17. _____

18. $-8 \div y$ 18. _____

19. $x + 11$ 19. _____

20. yz 20. _____

21. $x + y + z$ 21. _____

22. $7y$ 22. _____

23. $3z$ 23. _____

24. Graph the set of integers $\{-2, 1, -4\}$ on the number line.   24. _____

NAME _____ DATE _____ PERIOD _____

Test, Form 2B

SCORE _____

Write the letter for the correct answer in the blank at the right of each question.

1. Which integer represents a loss of $20?
   A. $|-20|$ B. -20 C. $|20|$ D. 20

   1. _____

2. What is the value of $|-4|$?
   F. -4 G. $-|-4|$ H. 4 I. 8

   2. _____

3. What is the value of $|7| + |-3|$?
   A. 10 B. 4 C. -4 D. -10

   3. _____

4. Which integers represent Q and R on the number line?
   F. $Q, -3; R, 1$ H. $Q, 3; R, -1$
   G. $Q, -1; R, 3$ I. $Q, 1; R, -3$

   4. _____

5. What value of s makes $-18 - (-5) = s$ a true sentence?
   A. 23 B. 13 C. -13 D. -23

   5. _____

6. A snake dives towards deeper water at a rate of 18 inches per second. If the snake continues at this rate for a total of 9 seconds, which expression represents this situation?
   F. $9 \div 18$ G. $9 \div (-18)$ H. $9(-18)$ I. $-9(-18)$

   6. _____

7. Amanda has $31 in her purse. She pays $6 for lunch. Which expression represents this situation?
   A. $31 + 6$ B. $31 + (-6)$ C. $-31 + 6$ D. $-31 + (-6)$

   7. _____

8. The highest point in Argentina is 6,960 meters above sea level. The lowest point is 105 meters below sea level. What is the different in meters between the highest and lowest elevation points?
   F. 7,065 m   G. 6,960 m   H. 6,855 m   I. 105 m

   8. _____

9. The lowest temperature on Mars can reach $-140°C$ while the highest temperature can be $20°C$. What is the difference between the highest and lowest temperatures on Mars?
   A. $120°C$ B. $140°C$ C. $160°C$ D. $200°C$

   9. _____

Course 2 • Chapter 3 Integers

NAME _____ DATE _____ PERIOD _____

Test, Form 2B   (continued)

SCORE _____

10. A window cleaner rises 36 feet up the side of a building, then descends 17 feet. What is his final position?

  F. 53 ft   G. 19 ft   H. −19 ft   I. −53 ft

10. _____

What is the value of each expression?

11. $11 + (-7)$
  A. −18   B. −4   C. 4   D. 18

11. _____

12. $-5(-11)$
  F. 55   G. 16   H. −16   I. −55

12. _____

13. $-15 \div (-3)$
  A. 18   B. 5   C. −5   D. −18

13. _____

14. $-39 - 7$
  F. −46   G. −32   H. 46   I. 273

14. _____

15. $8(-4)$
  A. 32   B. 4   C. −2   D. −32

15. _____

Evaluate each expression if $a = 4$, $b = 7$, and $c = -5$.

16. $b - (-4)$

16. _____

17. ac

17. _____

18. $\dfrac{-7}{b}$

18. _____

19. $a + 13$

19. _____

20. $a - b$

20. _____

21. $c + 6$

21. _____

22. $3c$

22. _____

23. $-8 \div a$

23. _____

24. Graph the set of integers {−1, 1, 4} on the number line.

24. ←—+—+—+—+—+—+—+—→

68   Course 2 · Chapter 3 Integers

NAME _____ DATE _____ PERIOD _____

Test, Form 3A

SCORE _____

Write an integer for each situation.

1. a deposit of $45

2. 5°C below 0

3. a loss of 15 yards

For Exercises 4–6, evaluate each expression.

4. $|-11|$

5. $|4| + |-6|$

6. $|-8| - |-3|$

7. Graph the set of integers $\{-3, 2, -5\}$ on a number line.

8. The temperature in Athens was 30°F. Five minutes later it was 20°F. What was the average change in temperature per minute as an integer?

9. The upper atmosphere of Neptune can get as cold as $-218°C$. The inner core of the planet can be as hot as $7{,}000°C$. What is the difference between the two temperature extremes?

10. What value of x makes $13 - (-9) = x$ a true sentence?

11. What value of n makes $-7(4) = n$ a true sentence?

12. What value of t makes $(-2) + (-11) = t$ a true sentence?

13. What value of w makes $-24 - (-13) = w$ a true sentence?

Evaluate each expression.

14. $-11 - (-3)$

15. $-25 + (-12)$

16. $8(-11)$

17. $40 \div (-8)$

Course 2 · Chapter 3 Integers

Test, Form 3A (continued)

18. $4 - (-2)$

19. $(-3)^2$

20. $\dfrac{-6}{-3}$

21. $-7(5)$

22. $-6(-4)$

23. Trent saved up $500 for summer vacation. If he spends $25 a week for eight weeks, how much money does he have left from his saving? Explain.

Evaluate each expression if $x = -4$, $y = 6$, and $z = -3$.

24. $15 - (-y)$

25. $20 \div x$

26. $7 + z$

27. $-2(3z)$

28. $\dfrac{xy}{12}$

29. $x + y$

30. xy

31. $x - z$

18. _____
19. _____
20. _____
21. _____
22. _____

23. _____

24. _____
25. _____
26. _____
27. _____
28. _____
29. _____
30. _____
31. _____

NAME _____ DATE _____ PERIOD _____

Test, Form 3B

SCORE _____

Write an integer for each situation.

1. 300 feet above sea level

2. a loss of $13

3. a $45 deposit

For Exercises 4–6, evaluate each expression.

4. $|-13|$

5. $|3| + |-5|$

6. $|-5| - |7 - 4|$

7. Graph the set of integers on a number line: $\{-6, 2, -3\}$.

8. The temperature in St. Paul was 45°F. Five minutes later it was 30°F. What was the average change in temperature per minute as an integer?

9. Tony had $360 in his savings account. His bank statement showed two transactions. A withdrawal as −$45 and a deposit as $25. What is the balance in his savings account after the two transactions?

10. What value of w makes $19 - (-10) = w$ a true sentence?

11. What value of c makes $-9(3) = c$ a true sentence?

12. What value of k makes $(-3) + (-14) = k$ a true sentence?

13. What value of d makes $-25 - (-16) = d$ a true sentence?

Evaluate each expression.

14. $-9 - (-13)$

15. $\dfrac{60}{-3}$

16. $6 - (-3)$

17. $(-5)^2$

1. _____
2. _____
3. _____
4. _____
5. _____
6. _____
7. ←++++++++++→
8. _____
9. _____
10. _____
11. _____
12. _____
13. _____
14. _____
15. _____
16. _____
17. _____

Course 2 · Chapter 3 Integers

Test, Form 3B (continued)

18. $-1 + (-11)$

19. $-10 - 7$

20. $-8 \div (-4)$

21. $-6(7)$

22. $-5(-3)$

23. Jean saved up $750 for summer vacation. If she spends $30 a week for nine weeks, how much money does she have left from her savings? Explain.

18. _____

19. _____

20. _____

21. _____

22. _____

23. _____

Evaluate each expression if $a = -8$, $b = 5$, and $c = -2$.

24. $8 - (-b)$

25. $16 \div c$

26. $9 + c$

27. $-5(3c)$

28. $ab \div 10$

29. $a + b$

30. ab

31. $a - c$

24. _____

25. _____

26. _____

27. _____

28. _____

29. _____

30. _____

31. _____

NAME _____ DATE _____ PERIOD _____

Are You Ready?

Review

> A fraction is in simplest form when there are no common factors of the numerator and denominator except 1.

Example 1
Write $\frac{25}{30}$ in simplest form.

Find the GCF of 25 and 30 by first finding the prime factorizations.

$25 = 5 \times 5$

$30 = 2 \times 3 \times 5$

The only factor they have in common is 5, which indicates it is the GCF.

$\frac{25}{30} = \frac{25 \div 5}{30 \div 5} = \frac{5}{6}$

Example 2
Write $\frac{24}{42}$ in simplest form.

Find the GCF of 24 and 42 by first finding the prime factorizations.

$24 = 2 \times 2 \times 2 \times 3$

$42 = 2 \times 3 \times 7$

The factors they have in common are 2 and 3. Therefore, 2×3 or 6 is the GCF.

$\frac{24}{42} = \frac{24 \div 6}{42 \div 6} = \frac{4}{7}$

Exercises

Write in simplest form.

1. $\frac{6}{8}$
2. $\frac{12}{15}$
3. $\frac{16}{24}$
4. $\frac{3}{6}$
5. $\frac{35}{50}$
6. $\frac{7}{7}$
7. $\frac{4}{12}$
8. $\frac{19}{20}$

1. _____
2. _____
3. _____
4. _____
5. _____
6. _____
7. _____
8. _____

Course 2 • Chapter 4 Rational Numbers

Are You Ready?

Practice

State which decimal is greater.

1. 0.7, 0.07
2. 0.36, 0.49
3. 0.29, 0.24

1. _____
2. _____
3. _____

Graph on a number line.

4. $2\frac{3}{4}$
5. $4\frac{1}{3}$

4.
5.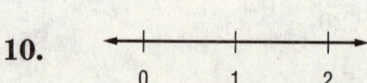

Write in simplest form.

6. $\frac{6}{24}$
7. $\frac{44}{66}$
8. $\frac{63}{70}$

6. _____
7. _____
8. _____

Solve each problem.

9. **PEANUTS** Snacks For Less sells one pound of peanuts for $3.57. Economy Nuts sells one pound for $3.52. At which store do peanuts cost less?

9. _____

10. **FISH** In an aquarium $\frac{2}{3}$ of the fish are gold and $\frac{1}{5}$ of the fish are black. Graph the numbers of gold and black fish on a number line.

10.

NAME _____ DATE _____ PERIOD _____

Are You Ready?

Apply

1. **SHOPPING** Marcia spent $0.76 for a cookie. Ken spent $0.67 for a candy bar. Who spent more?

2. **FISHING** Bruce's fish weighed 0.96 pound while Mosey's fish weighed 1.16 pounds. Whose fish weighed more?

3. **SHOES** Twenty of the 25 students in Mrs. Thigpen's class are wearing sneakers. What is this fraction in simplest form?

4. **FOOTBALL** Hill ran for 80 of his team's 120 rushing yards in Saturday's football game. Sheets ran for 60 of his team's 80 rushing yards. What are these fractions in simplest form?

5. **SOFTWARE** Leslie downloaded two software programs onto her computer. The first program took 5.76 minutes to download while the second took 5.06 minutes to download. Which program took longer to download?

6. **HIKING** Morgan stopped to rest after hiking 1.8 kilometers. Her mother rested after 2.1 kilometers. Who hiked farther?

Course 2 • Chapter 4 Rational Numbers

Diagnostic Test

State which decimal is greater.

1. 2.36, 2.63

2. 0.51, 0.051

1. _____

2. _____

Graph on a number line.

3. $\frac{1}{3}$

4. $\frac{4}{5}$

3.

4.

Write in simplest form.

5. $\frac{12}{14}$

6. $\frac{8}{20}$

7. $\frac{9}{10}$

8. $\frac{7}{28}$

5. _____

6. _____

7. _____

8. _____

Solve each problem.

9. **SPANISH** The school newspaper reported that 20 out of every 60 students in the school take Spanish. Write a fraction in simplest form for these results.

9. _____

10. **SPELLING** Shawn got 10 out of 20 spelling words correct while Natalie got 15 out of 20 spelling words correct. Graph their scores on a number line.

10.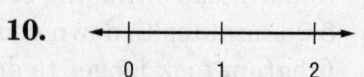

76   Course 2 • Chapter 4 Rational Numbers

Pretest

Write each fraction as a decimal.

1. $\frac{5}{8}$

2. $\frac{2}{11}$

1. _____

2. _____

Add, subtract, multiply, or divide. Write in simplest form.

3. $\frac{7}{10} - \frac{1}{10}$

4. $\frac{3}{5} + \frac{1}{2}$

5. $2\frac{3}{8} + 1\frac{1}{8}$

6. $5\frac{2}{5} - 3\frac{7}{10}$

7. $5\frac{1}{2} \times \frac{3}{11}$

8. $\frac{5}{6} \div \frac{1}{3}$

3. _____

4. _____

5. _____

6. _____

7. _____

8. _____

9. Replace ● with <, >, or = to make $-2\frac{1}{6}$ ● $-2\frac{5}{6}$ a true sentence.

9. _____

Complete. Round to the nearest hundredth if necessary.

10. 10 in. ≈ _____ cm

11. 1.4 lb ≈ _____ kg

12. 7 qt ≈ _____ mL

10. _____

11. _____

12. _____

13. **SOCCER** Paul hit the ball $\frac{1}{10}$ of the length of the field, and Mark hit the ball $\frac{3}{8}$ of the length of the field. Together, what fraction of the length of the field did they hit the ball?

13. _____

Course 2 • Chapter 4 Rational Numbers

NAME _____ DATE _____ PERIOD _____

Chapter Quiz

Write each fraction or mixed number as a decimal. Use bar notation if the decimal is a repeating decimal.

1. $2\frac{5}{8}$

2. $\frac{7}{9}$

3. $\frac{3}{4}$

1. _____

2. _____

3. _____

Replace each ● with <, >, or = to make a true sentence.

4. $\frac{24}{30}$ ● $\frac{45}{50}$

5. $\frac{6}{17}$ ● $\frac{18}{51}$

4. _____

5. _____

Add or subtract. Write in simplest form.

6. $\frac{5}{7} + \frac{3}{7}$

7. $\frac{7}{12} + \frac{3}{4}$

8. $-\frac{3}{5} - \frac{1}{5}$

9. $\frac{5}{8} - \frac{3}{16}$

6. _____

7. _____

8. _____

9. _____

10. **HOMEWORK** Honon spent $3\frac{1}{4}$ hours on homework yesterday while Sequoia spent $2\frac{5}{6}$ hours on homework. How much more time did Honon spend on homework than Sequoia?

10. _____

Course 2 • Chapter 4 Rational Numbers

NAME _____ DATE _____ PERIOD _____

Vocabulary Test

SCORE _____

bar notation	rational number
common denominator	repeating decimal
least common denominator (LCD)	terminating decimal
like fractions	unlike fractions

Write whether each sentence is *true* or *false*. If *false*, replace the underlined word or phrase to make a true sentence.

1. To add $\frac{4}{9}$ and $\frac{1}{9}$, first add <u>4 and 1</u>.   1. _____

2. To multiply fractions, multiply the numerators and <u>add</u> the denominators.   2. _____

3. To subtract like fractions, subtract the <u>numerators</u>.   3. _____

4. To divide $\frac{5}{8}$ by $\frac{1}{4}$, multiply $\frac{5}{8}$ by <u>4</u>.   4. _____

5. A <u>repeating decimal</u> is a decimal that ends when it reaches a remainder of zero.   5. _____

6. <u>Bar notation</u> is used to indicate that a number pattern repeats indefinitely.   6. _____

7. To multiply powers with the same base, <u>multiply</u> the exponents.   7. _____

8. To <u>subtract</u> powers with the same base, subtract the exponents.   8. _____

Define each term in your own words.

9. least common denominator   9. _____

10. rational number   10. _____

Standardized Test Practice

Read each question. Then fill in the correct answer on the answer sheet provided by your teacher or on a sheet of paper.

1. **SHORT RESPONSE** Mrs. Brown needs to make two different desserts for a party. The first recipe requires $2\frac{1}{4}$ cups of flour and the second recipe requires $\frac{3}{4}$ cup less than the first. Write an equation that can be used to find the number of cups of flour needed for the second recipe.

2. The fraction $\frac{5}{6}$ is found between which pair of fractions on a number line?
   A. $\frac{1}{4}$ and $\frac{5}{8}$
   B. $\frac{1}{3}$ and $\frac{4}{9}$
   C. $\frac{11}{12}$ and $\frac{31}{36}$
   D. $\frac{7}{12}$ and $\frac{17}{18}$

3. At 7 A.M., the temperature was 15°F below zero. By 2 P.M. the temperature rose 32°F and by 5 P.M. it dropped 10°F. What was the temperature at 5 P.M.?
   F. 10°F
   G. 9°F
   H. 7°F
   I. 11°F

4. **GRIDDED RESPONSE** A diver is swimming 11 meters below the surface. The diver sees a shark 19 meters below him. How many meters below the surface is the shark?

5. **Gridded Response** Maria had $240 in her savings account. The table shows the change in her account for four consecutive weeks.

Week	Change
1	Deposit of $25
2	Withdrawal of $45
3	Withdrawal of $10
4	Deposit of $60

   How much money, in dollars, did Maria have in her account at the end of the four weeks?

6. The table shows the distance Kelly swam over a four-day period. What was the total distance, in miles, that Kelly swam?

Kelly's Swimming	
Day	Distance (mi)
Monday	1.5
Tuesday	$2\frac{3}{4}$
Wednesday	2.3
Thursday	$3\frac{1}{2}$

   A. 10.5 miles
   B. $10\frac{1}{4}$ miles
   C. $10\frac{1}{20}$ miles
   D. 9 miles

7. Which of the following gives the correct meaning of the expression $\frac{5}{8} \div \frac{1}{3}$?
   F. $\frac{5}{8} \div \frac{1}{3} = \frac{8}{5} \times \frac{3}{1}$
   G. $\frac{5}{8} \div \frac{1}{3} = \frac{5+1}{8+3}$
   H. $\frac{5}{8} \div \frac{1}{3} = \frac{5}{8} \times \frac{3}{1}$
   I. $\frac{5}{8} \div \frac{1}{3} = \frac{5}{8} \times \frac{1}{3}$

Course 2 • Chapter 4 Rational Numbers

8. The table shows the lowest temperature readings to the nearest degree recorded for four countries.

City	Temperature (°F)
Finland	−61°
France	−42°
India	−27°
United States	−80°

Which of the countries has the lowest recorded temperature?

A. Finland
B. India
C. France
D. United States

9. **GRIDDED RESPONSE** Nate had 25 action figures. He gave away 10 to his brother. He then got 3 new action figures as a gift. How many action figures does Nate have now?

10. Which expression represents the least value?

F. $678 \div \frac{1}{3}$

G. $678 + \frac{1}{3}$

H. $678 \times \frac{1}{3}$

I. $678 - \frac{1}{3}$

11. **GRIDDED RESPONSE** Jacob had $25 for back-to-school shopping. He bought a shirt for $15 and then returned a shirt he bought a week ago and got $20 in return. How much money in dollars does Jacob have now?

12. **GRIDDED RESPONSE** Evan runs $2\frac{3}{8}$ miles each week. He runs $\frac{3}{4}$ mile on Mondays and $\frac{3}{4}$ mile on Tuesdays. How far does he run, in miles, on Thursday if it is the only other day he runs?

13. **SHORT RESPONSE** A recipe for a batch of cookies calls for $2\frac{1}{3}$ cups of flour for 24 cookies. Manuel wants to make 72 cookies. How many cups of flour will he need?

14. **EXTENDED RESPONSE** A box of laundry detergent contains 35 cups. It takes $1\frac{1}{4}$ cups per load of laundry.

Part A Write an equation to represent how many loads ℓ you can wash with one box.

Part B How many loads can you wash with one box?

Part C How many loads can you wash with 3 boxes?

Course 2 • Chapter 4 Rational Numbers

NAME _____ DATE _____ PERIOD _____

SCORE _____

Student Recording Sheet

Use this recording sheet with the Standardized Test Practice pages.

Fill in the correct answer. For gridded-response questions, write your answers in the boxes on the answer grid and fill in the bubbles to match your answers.

1. _____

2. Ⓐ Ⓑ Ⓒ Ⓓ

3. Ⓕ Ⓖ Ⓗ Ⓘ

4. [gridded response]

5. [gridded response]

6. Ⓐ Ⓑ Ⓒ Ⓓ

7. Ⓕ Ⓖ Ⓗ Ⓘ

8. Ⓐ Ⓑ Ⓒ Ⓓ

9. [gridded response]

10. Ⓕ Ⓖ Ⓗ Ⓘ

11. [gridded response]

12. [gridded response]

13. _____

Extended Response
Record your answers for Exercise 14 on the back of this paper.

82   Course 2 • Chapter 4 Rational Numbers

NAME _____ DATE _____ PERIOD _____

Extended-Response Test

SCORE _____

Demonstrate your knowledge by giving a clear, concise solution to each problem. Be sure to include all relevant drawings and justify your answers. You may show your solutions in more than one way or investigate beyond the requirements of the problem. If necessary, record your answer on another piece of paper.

1. Jennifer is shopping for food for her party.

   a. She bought $1\frac{1}{2}$ pounds of bologna and $2\frac{1}{4}$ pounds of turkey breast. How much more turkey breast than bologna did she buy? Find the answer using two different methods.

   b. Draw a model that illustrates $2\frac{1}{2} \times \frac{1}{2}$.

   c. If a recipe for punch calls for $\frac{1}{2}$ gallon of fruit drink, how much fruit drink will Jennifer need to make $2\frac{1}{2}$ batches of the recipe?

2. **a.** Write a word problem that uses mixed numbers and asks to find the unit price of an item.

   b. Find the unit price for part a.

3. One brand of popcorn costs $2.00 for 1 pound 9 ounces and the second brand costs $2.10 for 1 pound 14 ounces. Which is the better buy? Why? (*Hint:* There are 16 ounces in one pound. What is the cost per ounce?)

Course 2 • Chapter 4 Rational Numbers

NAME _____ DATE _____ PERIOD _____

Extended-Response Rubric

SCORE _____

Score	Description
4	A score of four is a response in which the student demonstrates a thorough understanding of the mathematics concepts and/or procedures embodied in the task. The student has responded correctly to the task, used mathematically sound procedures, and provided clear and complete explanations and interpretations. The response may contain minor flaws that do not detract from the demonstration of a thorough understanding.
3	A score of three is a response in which the student demonstrates an understanding of the mathematics concepts and/or procedures embodied in the task. The student's response to the task is essentially correct with the mathematical procedures used and the explanations and interpretations provided demonstrating an essential but less than thorough understanding. The response may contain minor flaws that reflect inattentive execution of mathematical procedures or indications of some misunderstanding of the underlying mathematics concepts and/or procedures.
2	A score of two indicates that the student has demonstrated only a partial understanding of the mathematics concepts and/or procedures embodied in the task. Although the student may have used the correct approach to obtaining a solution or may have provided a correct solution, the student's work lacks an essential understanding of the underlying mathematical concepts. The response contains errors related to misunderstanding important aspects of the task, misuse of mathematical procedures, or faulty interpretations of results.
1	A score of one indicates that the student has demonstrated a very limited understanding of the mathematics concepts and/or procedures embodied in the task. The student's response is incomplete and exhibits many flaws. Although the student's response has addressed some of the conditions of the task, the student reached an inadequate conclusion and/or provided reasoning that was faulty or incomplete. The response exhibits many flaws or may be incomplete.
0	A score of zero indicates that the student has provided no response at all, or a completely incorrect or uninterpretable response, or demonstrated insufficient understanding of the mathematics concepts and/or procedures embodied in the task. For example, a student may provide some work that is mathematically correct, but the work does not demonstrate even a rudimentary understanding of the primary focus of the task.

NAME _____ DATE _____ PERIOD _____

Test, Form 1A

SCORE _____

Write the letter for the correct answer in the blank at the right of each question.

1. What is $\frac{4}{5}$ as a decimal?
   A. 8.0   B. $0.\overline{8}$ C. 0.8   D. 0.08

   1. _____

2. What is $1\frac{8}{9}$ as a decimal?
   F. $0.1\overline{8}$ G. 1.8   H. $1.\overline{8}$ I. 18.9

   2. _____

3. What is 0.125 as a fraction in simplest form?
   A. $\frac{1}{8}$ B. $\frac{1}{6}$ C. $\frac{1}{5}$ D. $\frac{3}{8}$

   3. _____

4. What is the LCD of $\frac{11}{12}$ and $\frac{1}{8}$?
   F. 96   G. 88   H. 24   I. 12

   4. _____

5. Which symbol makes $3\frac{13}{28}$ ● $3\frac{17}{30}$ a true sentence?
   A. >   B. <   C. =   D. ×

   5. _____

6. Which symbol makes $-\frac{7}{12}$ ● $-\frac{5}{9}$ a true sentence?
   F. >   G. <   H. =   I. ÷

   6. _____

7. Kelly made six batches of cookies. Her friends asked her to make an extra batch. She used $17\frac{1}{4}$ cups of sugar and needs another $\frac{7}{8}$ cups for the extra batch. How many cups of sugar will Kelly use altogether?
   A. $16\frac{3}{8}$ B. $\frac{132}{8}$ C. $18\frac{1}{8}$ D. $18\frac{1}{4}$

   7. _____

For Exercises 8–13, what is the value of each expression in simplest form?

8. $\frac{1}{6} + \left(-\frac{7}{9}\right)$
   F. $-\frac{7}{54}$ G. $\frac{8}{15}$ H. $-\frac{11}{18}$ I. $1\frac{1}{18}$

   8. _____

9. $11\frac{5}{7} + 8\frac{1}{7}$
   A. $18\frac{6}{7}$ B. $19\frac{6}{7}$ C. 20   D. $20\frac{1}{7}$

   9. _____

10. $\frac{3}{12} \times \frac{4}{21}$
    F. $\frac{7}{12}$ G. $\frac{1}{3}$ H. $\frac{12}{252}$ I. $\frac{1}{21}$

    10. _____

Course 2 • Chapter 4 Rational Numbers

Test, Form 1A (continued)

11. $\frac{3}{5} - \frac{1}{5}$

    A. $\frac{2}{25}$ B. $\frac{1}{5}$ C. $\frac{2}{5}$ D. $\frac{4}{5}$ 11. _____

12. $3\frac{5}{9} - 2\frac{1}{3}$

    F. $\frac{2}{9}$ G. $1\frac{4}{27}$ H. $1\frac{2}{9}$ I. $1\frac{2}{3}$ 12. _____

13. $\frac{5}{6} \div 2\frac{1}{6}$

    A. $\frac{5}{13}$ B. $\frac{5}{12}$ C. $1\frac{29}{36}$ D. $2\frac{3}{13}$ 13. _____

14. It takes Lara $\frac{1}{3}$ hour to walk to the library $\frac{3}{4}$ mile away. What is her walking pace?

    F. $\frac{1}{4}$ mile per hour   H. $2\frac{1}{4}$ miles per hour

    G. $\frac{4}{9}$ mile per hour   I. 15 miles per hour 14. _____

15. A recipe calls for $\frac{1}{8}$ teaspoon of vanilla extract. If the recipe is doubled, how much vanilla extract is needed?

    A. $\frac{1}{16}$ tsp B. $\frac{1}{4}$ tsp C. $\frac{1}{2}$ tsp D. 1 tsp 15. _____

16. Jeremy and his friends ate $\frac{5}{8}$ of a pie. If the pie was cut into eight pieces, how much pie is left over?

    F. $\frac{1}{8}$ G. $\frac{2}{8}$ H. $\frac{1}{4}$ I. $\frac{3}{8}$ 16. _____

Complete. Round to the nearest hundredth if necessary.

17. 927.5 g ≈ _____ lb

    A. 420.71 B. 2.04 C. 148.32 D. 57.97 17. _____

18. 18 qt ≈ _____ mL

    F. 1.902 G. 17.03 H. 19.02 I. 17,034.3 18. _____

19. 148 mL ≈ _____ c

    A. 6.3 B. 15.99 C. 0.63 D. 1.60 19. _____

NAME _____ DATE _____ PERIOD _____

Test, Form 1B

SCORE _____

Write the letter for the correct answer in the blank at the right of each question.

1. What is $\frac{3}{5}$ as a decimal?
   A. 0.06 B. 0.6 C. $0.\overline{6}$ D. 6

   1. _____

2. What is $2\frac{4}{9}$ as a decimal?
   F. $2.0\overline{4}$ G. $2.\overline{4}$ H. $2.\overline{49}$ I. $2.\overline{5}$

   2. _____

3. What is 0.375 as a fraction in simplest form?
   A. $\frac{1}{8}$ B. $\frac{3}{8}$ C. $\frac{5}{16}$ D. $\frac{2}{5}$

   3. _____

4. What is the LCD of $\frac{11}{15}$ and $\frac{5}{6}$?
   F. 90 G. 60 H. 30 I. 3

   4. _____

5. Which symbol makes $6\frac{15}{28}$ ● $6\frac{5}{9}$ a true sentence?
   A. < B. > C. = D. ×

   5. _____

6. Which symbol makes $-\frac{3}{4}$ ● $-\frac{11}{12}$ a true sentence?
   F. < G. > H. = I. ÷

   6. _____

7. Mario made six batches of cookies. His friends asked him to make an extra batch. He used $15\frac{3}{4}$ cups of sugar and needs another $\frac{5}{8}$ cups for the extra batch. How many cups of sugar will Mario use altogether?
   A. $17\frac{1}{4}$ B. $16\frac{3}{8}$ C. $\frac{128}{8}$ D. $15\frac{1}{8}$

   7. _____

For Exercises 8–13, what is the value of each expression in simplest form?

8. $\frac{1}{8} + \left(-\frac{3}{16}\right)$
   F. $\frac{4}{24}$ G. $\frac{5}{16}$ H. $-\frac{5}{16}$ I. $-\frac{1}{16}$

   8. _____

9. $8\frac{2}{7} + 10\frac{4}{7}$
   A. $18\frac{3}{7}$ B. $18\frac{8}{49}$ C. $18\frac{6}{7}$ D. $18\frac{8}{7}$

   9. _____

10. $\frac{2}{21} \times \frac{7}{5}$
    F. $\frac{7}{13}$ G. $\frac{9}{26}$ H. $\frac{3}{35}$ I. $\frac{2}{15}$

    10. _____

Course 2 • Chapter 4 Rational Numbers

Test, Form 1B (continued)

11. $\frac{7}{8} - \frac{3}{8}$
    A. $\frac{1}{2}$
    B. $\frac{4}{0}$
    C. $\frac{1}{16}$
    D. $\frac{5}{4}$

    11. _____

12. $4\frac{7}{9} - 3\frac{2}{3}$
    F. $1\frac{5}{6}$
    G. $1\frac{1}{3}$
    H. $1\frac{1}{9}$
    I. $\frac{8}{9}$

    12. _____

13. $\frac{3}{4} \div 4\frac{1}{4}$
    A. $\frac{3}{17}$
    B. $3\frac{1}{4}$
    C. $3\frac{3}{16}$
    D. $\frac{17}{3}$

    13. _____

14. Michaela walked $\frac{2}{5}$ mile to the store in $\frac{1}{4}$ hour. What is her walking pace?
    F. $\frac{1}{10}$ mile per hour
    G. $\frac{5}{8}$ mile per hour
    H. $1\frac{3}{5}$ miles per hour
    I. $2\frac{1}{2}$ miles per hour

    14. _____

15. Tameron made a recipe that called for $\frac{1}{4}$ teaspoon of salt. If the recipe is doubled, how much salt is needed?
    A. $\frac{1}{12}$ tsp
    B. $\frac{1}{8}$ tsp
    C. $\frac{1}{4}$ tsp
    D. $\frac{1}{2}$ tsp

    15. _____

16. Eliot and his friends ate $\frac{3}{8}$ of a pie. If the pie was cut into eight pieces, how much pie is left over?
    F. $\frac{5}{8}$
    G. $\frac{2}{8}$
    H. $\frac{1}{4}$
    I. $\frac{3}{8}$

    16. _____

Complete. Round to the nearest hundredth if necessary.

17. 37 mL ≈ _____ fl oz
    A. 29.57
    B. 6.39
    C. 0.80
    D. 1.25

    17. _____

18. 47 kg ≈ _____ lb
    F. 103.62
    G. 103.4
    H. 21.32
    I. 21.37

    18. _____

19. 78 ft ≈ _____ m
    A. 260
    B. 85.71
    C. 23.4
    D. 70.98

    19. _____

NAME _____ DATE _____ PERIOD _____

Test, Form 2A

SCORE _____

Write the letter for the correct answer in the blank at the right of each question.

1. What is $4\frac{2}{3}$ as a decimal?
   A. $0.4\overline{6}$ B. 4.6 C. $4.\overline{6}$ D. $46.\overline{6}$

   1. _____

2. What is 0.82 as a fraction in simplest form?
   F. $\frac{41}{50}$ G. $\frac{8}{10}$ H. $\frac{12}{25}$ I. $\frac{2}{5}$

   2. _____

3. What is the LCD of $\frac{11}{7}$ and $\frac{3}{8}$?
   A. 14   B. 16   C. 33   D. 56

   3. _____

4. Which symbol makes $\frac{7}{11}$ ● $\frac{3}{5}$ a true sentence?
   F. $>$ G. $<$ H. $=$ I. $+$

   4. _____

5. Which of the following has the least value?
   A. $\frac{13}{15}$ B. $\frac{7}{8}$ C. $\frac{2}{3}$ D. $\frac{3}{4}$

   5. _____

For Exercises 6–13, what is the value of each expression in simplest form?

6. $-\frac{11}{15} + \left(-\frac{2}{15}\right)$
   F. $-\frac{9}{15}$ G. $-\frac{13}{15}$ H. $\frac{13}{15}$ I. $3\frac{1}{3}$

   6. _____

7. $15\frac{3}{5} - 3\frac{2}{7}$
   A. $12\frac{31}{35}$ B. $12\frac{11}{35}$ C. $11\frac{21}{35}$ D. $11\frac{1}{35}$

   7. _____

8. $\frac{2}{3} + \frac{3}{8}$
   F. $\frac{1}{6}$ G. $\frac{5}{24}$ H. $\frac{5}{11}$ I. $1\frac{1}{24}$

   8. _____

9. $2\frac{1}{8} + 1\frac{5}{12}$
   A. $3\frac{13}{24}$ B. $3\frac{3}{10}$ C. $3\frac{1}{4}$ D. $3\frac{5}{96}$

   9. _____

10. $2\frac{5}{6} \times \frac{1}{3}$
    F. $3\frac{1}{6}$ G. $1\frac{8}{9}$ H. $\frac{17}{9}$ I. $\frac{17}{18}$

    10. _____

Course 2 • Chapter 4 Rational Numbers

NAME _____ DATE _____ PERIOD _____

Test, Form 2A (continued)   SCORE _____

11. $\frac{9}{10} - \frac{3}{10}$

   A. $\frac{3}{50}$ B. $\frac{1}{2}$ C. $\frac{3}{5}$ D. $\frac{6}{5}$ 11. _____

12. $10\frac{1}{6} - 4\frac{3}{14}$

   F. $5\frac{20}{21}$ G. $5\frac{1}{4}$ H. $6\frac{1}{4}$ I. $6\frac{20}{21}$ 12. _____

13. $4\frac{1}{4} \div 2\frac{1}{2}$

   A. $1\frac{5}{8}$ B. $1\frac{7}{10}$ C. $1\frac{3}{4}$ D. $6\frac{3}{4}$ 13. _____

14. Usually Ellis rides his bicycle $5\frac{4}{5}$ miles a day. Today he rode half his usual distance. How far did he ride?   14. _____

15. A recipe calls for $1\frac{3}{4}$ cups of flour. If the recipe is tripled, how much flour is needed?   15. _____

16. Find the perimeter of the figure.   16. _____

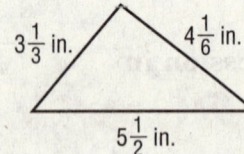

17. Find the area of a rectangle with a length of 3 feet and a width of $2\frac{7}{8}$ feet.   17. _____

18. Ayana bought a container of peanuts. She gave $\frac{1}{4}$ of it to one sister, $\frac{1}{3}$ to another sister, and she kept the rest for herself. What fraction did she keep?   18. _____

Complete. Round to the nearest hundredth if necessary.

19. 17 km ≈ _____ mi   19. _____

20. 1,450 kg ≈ _____ T   20. _____

21. 2.3 pt ≈ _____ mL   21. _____

22. 72 g ≈ _____ lb   22. _____

23. 0.53 c ≈ _____ mL   23. _____

24. 8 m ≈ _____ ft   24. _____

NAME _____ DATE _____ PERIOD _____

Test, Form 2B

SCORE _____

Write the letter for the correct answer in the blank at the right of each question.

1. What is $3\frac{4}{9}$ as a decimal?
   A. $0.3\overline{4}$ B. 3.4 C. $3.\overline{4}$ D. $34.\overline{4}$

   1. _____

2. What is 0.75 as a fraction in simplest form?
   F. $\frac{7}{5}$ G. $\frac{75}{100}$ H. $\frac{3}{4}$ I. $\frac{1}{2}$

   2. _____

3. What is the LCD of $\frac{1}{4}$ and $\frac{2}{3}$?
   A. 4 B. 7 C. 12 D. 18

   3. _____

4. Which symbol makes $\frac{4}{5}$ ● $\frac{2}{3}$ a true sentence?
   F. > G. < H. = I. +

   4. _____

5. Which of the following has the least value?
   A. $\frac{5}{7}$ B. $\frac{7}{8}$ C. $\frac{11}{15}$ D. $\frac{7}{10}$

   5. _____

For Exercises 6–13, what is the value of each expression in simplest form?

6. $-\frac{9}{14} + \left(-\frac{3}{14}\right)$
   F. $-\frac{3}{7}$ G. $-\frac{6}{7}$ H. $-\frac{7}{6}$ I. $1\frac{1}{7}$

   6. _____

7. $14\frac{1}{5} - 5\frac{5}{6}$
   A. $8\frac{11}{30}$ B. $9\frac{4}{30}$ C. $11\frac{4}{5}$ D. $19\frac{1}{30}$

   7. _____

8. $\frac{4}{5} + \frac{5}{6}$
   F. $1\frac{2}{3}$ G. $1\frac{19}{30}$ H. $\frac{49}{30}$ I. 9

   8. _____

9. $1\frac{2}{3} + 2\frac{3}{4}$
   A. $3\frac{5}{12}$ B. $4\frac{1}{4}$ C. $4\frac{1}{3}$ D. $4\frac{5}{12}$

   9. _____

10. $1\frac{1}{2} \times \frac{3}{4}$
    F. $\frac{7}{8}$ G. 1 H. $1\frac{1}{8}$ I. 2

    10. _____

Course 2 • Chapter 4 Rational Numbers

Test, Form 2B (continued)

11. $\frac{7}{8} - \frac{3}{8}$

    A. $\frac{1}{4}$ B. $\frac{1}{2}$ C. $\frac{5}{8}$ D. $\frac{5}{4}$

    11. _____

12. $9\frac{2}{3} - 7\frac{5}{6}$

    F. $1\frac{1}{6}$ G. $1\frac{5}{6}$ H. $2\frac{1}{6}$ I. $2\frac{5}{6}$

    12. _____

12. $2\frac{1}{2} \div 4\frac{3}{4}$

    A. $\frac{10}{19}$ B. $\frac{3}{5}$ C. $1\frac{9}{10}$ D. $\frac{95}{8}$

    13. _____

14. Usually Clarissa jogs $2\frac{3}{5}$ miles a day. Today she jogged half her usual distance. How far did she jog?

    14. _____

15. A recipe calls for $3\frac{3}{4}$ cups of milk. If the recipe is doubled, how much milk is needed?

    15. _____

16. Find the perimeter of the figure.

    16. _____

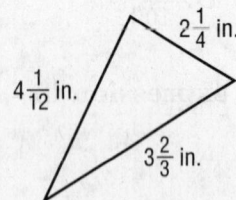

17. Find the area of a rectangle with a length of 5 inches and a width of $3\frac{5}{6}$ inches.

    17. _____

18. Chad bought a box of raisins. He gave $\frac{1}{6}$ of it to one brother, $\frac{1}{4}$ to another brother, and he kept the rest for himself. What fraction did he keep?

    18. _____

Complete. Round to the nearest hundredth if necessary.

19. 8.4 cm ≈ _____ in.

    19. _____

20. 43 km ≈ _____ mi

    20. _____

21. 1.2 lb ≈ _____ kg

    21. _____

22. 3 T ≈ _____ kg

    22. _____

23. 100 mL ≈ _____ c

    23. _____

24. 8 L ≈ _____ gal

    24. _____

NAME _____ DATE _____ PERIOD ____

Test, Form 3A

SCORE _____

Write each fraction or mixed number as a decimal. Use bar notation if the decimal is a repeating decimal.

1. $\dfrac{8}{9}$

2. $-6\dfrac{3}{4}$

Write each decimal as a fraction or mixed number in simplest form.

3. -0.02

4. 68.25

For Exercises 5 and 6, replace each ● with <, >, or = to make a true sentence.

5. $\dfrac{8}{13}$ ● $\dfrac{5}{17}$

6. $-\dfrac{10}{15}$ ● $-\dfrac{5}{14}$

7. Order the lengths $\dfrac{1}{4}$ inch, 0.5 inch, and $\dfrac{10}{25}$ inch from least to greatest.

For Exercises 8–15, add, subtract, multiply, or divide. Write in simplest form.

8. $-\dfrac{2}{9} - \dfrac{4}{9}$

9. $\dfrac{3}{8} \times \left(-\dfrac{2}{7}\right)$

10. $-\dfrac{7}{8} \div \dfrac{5}{6}$

11. $3\dfrac{4}{5} - 2\dfrac{1}{3}$

12. $2 + 7\dfrac{11}{12}$

13. $-\dfrac{2}{3} + \dfrac{4}{9}$

14. $3\dfrac{3}{5} \div 1\dfrac{2}{7}$

15. $\dfrac{4}{9} \times 9\dfrac{9}{16}$

1. _____
2. _____
3. _____
4. _____
5. _____
6. _____
7. _____
8. _____
9. _____
10. _____
11. _____
12. _____
13. _____
14. _____
15. _____

Course 2 • Chapter 4 Rational Numbers

Test, Form 3A (continued)

16. It takes Mara 50 minutes to walk to her friend's house $1\frac{2}{3}$ miles away. What is her walking pace in miles per hour?

16. _____

17. A recipe calls for $2\frac{3}{4}$ cups of milk. If the recipe is tripled, how much milk is needed?

17. _____

18. Tom practiced piano $1\frac{1}{3}$ hours on Monday and $\frac{5}{6}$ hour on Tuesday. How much did he practice in all those two days?

18. _____

19. Find the perimeter of the figure.

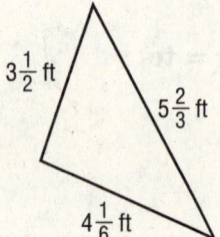

19. _____

20. A restaurant had 3 pies, each cut into eighths. By noon, $\frac{3}{4}$ of all the pieces were sold. How many pieces of pie were sold by noon?

20. _____

21. The Davis family traveled 35 miles in $\frac{1}{2}$ hour. If it is currently 2:00 P.M. and the family's destination is 245 miles away, at what time will they arrive? Explain how you solved the problem.

21. _____

Complete. Round to the nearest hundredth if necessary.

22. 21 ft ≈ _____ m

22. _____

23. 34.6 cm ≈ _____ in.

23. _____

24. 7.4 kg ≈ _____ lb

24. _____

25. 4.9 T ≈ _____ kg

25. _____

Test, Form 3B

Write each fraction or mixed number as a decimal. Use bar notation if the decimal is a repeating decimal.

1. $\dfrac{3}{11}$

2. $-7\dfrac{3}{5}$

Write each decimal as a fraction or mixed number in simplest form.

3. -0.06

4. 86.75

For Exercises 5 and 6, replace each ● with <, >, or = to make a true sentence.

5. $\dfrac{7}{9}$ ● $\dfrac{15}{17}$

6. $-\dfrac{7}{8}$ ● $-\dfrac{8}{10}$

7. Order the lengths $\dfrac{7}{8}$ feet, 0.8 feet, and $\dfrac{15}{16}$ feet from least to greatest.

For Exercises 8–15, add, subtract, multiply, or divide. Write in simplest form.

8. $-\dfrac{1}{14} - \dfrac{9}{14}$

9. $\dfrac{5}{9} \times \left(-\dfrac{3}{7}\right)$

10. $-\dfrac{1}{3} \div \dfrac{4}{9}$

11. $7\dfrac{5}{12} - 2\dfrac{2}{3}$

12. $3 + 1\dfrac{2}{5}$

13. $-\dfrac{5}{7} + \dfrac{1}{5}$

14. $4\dfrac{3}{8} \div 1\dfrac{3}{4}$

15. $\dfrac{5}{6} \times 12\dfrac{3}{5}$

Course 2 • Chapter 4 Rational Numbers

Test, Form 3B (continued)

16. It takes Sarah 50 minutes to walk to her friend's house $1\frac{3}{4}$ miles away. What is her walking pace in miles per hour?

    16. _____

17. A recipe calls for $2\frac{1}{4}$ teaspoons of baking soda. If the recipe is tripled, how much baking soda is needed?

    17. _____

18. Ron practiced piano $1\frac{2}{3}$ hours on Monday and $\frac{5}{6}$ hour on Tuesday. How much did he practice in all those two days?

    18. _____

19. Find the perimeter of the figure.

    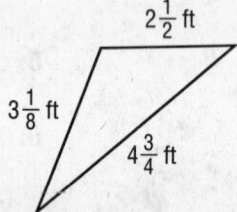

    19. _____

20. A restaurant had 5 pies, each cut into eighths. By noon, $\frac{3}{4}$ of all the pieces were sold. How many pieces of pie were sold by noon?

    20. _____

21. The Woods family traveled 25 miles in $\frac{1}{2}$ hour. If it is currently 3:00 P.M. and the family's destination is 225 miles away, at what time will they arrive? Explain how you solved the problem.

    21. _____

Complete. Round to the nearest hundredth if necessary.

22. 4.2 m ≈ _____ ft

    22. _____

23. 8.34 mi ≈ _____ km

    23. _____

24. 7.2 lb ≈ _____ kg

    24. _____

25. 1 L ≈ _____ gal

    25. _____

Are You Ready?

Review

Example 1

Find 8.4 × 3.

  8.4 ← 1 decimal place
× 3
25.2 ← 1 decimal place

So, 8.4 × 3 = 25.2.

Example 2

Find 2.6 × 5.1.

  2.6 ← 1 decimal place
× 5.1 ← 1 decimal place
  26
1300
13.26 ← 1 + 1 or 2 decimal places

So, 2.6 × 5.1 = 13.26.

Multiply.

1. 2 × 4.3

2. 7 × 8.1

3. 6.5 × 3

4. 9.6 × 4

5. 3.4 × 2.1

6. 1.1 × 3.2

7. 2.9 × 2.3

8. 7.6 × 1.8

9. 6.7 × 3.9

10. 8.5 × 7.6

1. _____

2. _____

3. _____

4. _____

5. _____

6. _____

7. _____

8. _____

9. _____

10. _____

Course 2 • Chapter 5 Expressions

NAME _____ DATE _____ PERIOD _____

Are You Ready?

Practice

Add or subtract.

1. 26.4 + 17.1

2. 9.6 + 42.5

3. 38.9 − 25.2

4. 75.64 − 50.73

5. 49.5 + 11.02

6. 86.79 − 34.6

7. **SHOPPING** Carlos bought the items shown in the table. What was the total cost, not including tax?

Item	Cost ($)
inline skates	68.99
knee pads	15.85

8. **JOGGING** Genevieve jogged 3.8 miles. How much farther does she have to jog to reach her goal of 5 miles?

Multiply or divide.

9. 8.6 × 2

10. 3 × 7.4

11. 35.88 ÷ 2.6

12. 6.65 ÷ 3.5

13. 9.2 × 4.1

14. 10.26 ÷ 1.8

15. **FOOD** Six bagels cost $10.50. How much does each bagel cost?

1. _____
2. _____
3. _____
4. _____
5. _____
6. _____
7. _____
8. _____
9. _____
10. _____
11. _____
12. _____
13. _____
14. _____
15. _____

NAME _____ DATE _____ PERIOD _____

Are You Ready?

Apply

1. **FOOD** Jasmine bought the items shown in the table. What was the total cost?

Item	Cost ($)
burrito	2.65
lemonade	1.70

2. **MONEY** Ethan had $15. He spent $13.74 on art supplies. How much money did Ethan have left?

3. **ENTERTAINMENT** Three friends spent $26.25 on movie tickets. How much does each ticket cost?

4. **PUPPY** The weight of a puppy is shown in the table. How much more did the puppy weigh in Week 8 than in Week 4?

Week	Weight (lb)
4	5.8
8	13.4

5. **JOBS** Ariana earns $12.35 per hour. What does she earn in one week if she works 18 hours?

6. **NUTRITION** A can of peas contains 3.5 servings and one serving is 0.5 cup. How many cups of peas are in the can?

Course 2 • Chapter 5 Expressions

NAME _____ DATE _____ PERIOD _____

Diagnostic Test

Add or subtract.

1. $12.7 + 9.4$

2. $54.16 + 10.3$

3. $62.5 - 14.2$

4. $43.8 + 22.17$

5. $80.6 - 67.41$

6. **SHOPPING** Brayden bought supplies for his camping trip. He gave the cashier $45.00 and received $16.02 in change. How much did Brayden spend on camping supplies?

7. **EXERCISE** The miles that Isabella walked on the treadmill are shown in the table. What is the total number of miles that Isabella walked on Wednesday and Thursday?

Day	Miles
Wednesday	1.16
Thursday	2.09

Multiply or divide.

8. 3.7×2

9. 8×1.5

10. $17.92 \div 2.8$

11. 12.3×7.4

12. $19.6 \div 5.6$

13. **MUSIC** Callie listened to nine songs on her MP3 player, and it took 34.2 minutes. What was the average length of each song?

14. **SOUP** A can of condensed chicken noodle soup is 10.75 ounces. It contains 2.5 servings. How many ounces of condensed soup is one serving?

15. **PARTY** It costs $8.95 to rent a round table for an outdoor party. The party planner estimates that 12 tables will be needed. How much will it cost to rent the tables?

1. _____
2. _____
3. _____
4. _____
5. _____
6. _____
7. _____
8. _____
9. _____
10. _____
11. _____
12. _____
13. _____
14. _____
15. _____

NAME _____ DATE _____ PERIOD _____

Pretest

Evaluate each expression if $a = 2$ and $b = 8$.

1. $3a + 1$

2. $\dfrac{24}{b}$

Describe the relationship between the terms in the arithmetic sequence. Then write the next three terms in each sequence.

3. 4, 9, 14, 19, …

4. 0, 14, 28, 42, …

5. **FISH** The table shows the cost of goldfish. How much will 6 goldfish cost?

Number of Fish	Cost ($)
1	2.80
2	5.60
3	8.40
4	11.20

Name the property shown by each statement.

6. $4m \cdot 0 \cdot 3m = 0$

7. $5 + (a + 17) = (5 + a) + 17$

Use the Distributive Property to rewrite each expression.

8. $4(x + 7)$

9. $-5(y + 10)$

10. Write $3x - 1 + 5x + 7$ in simplest form.

11. Find $(x + 1) + (x + 1)$.

12. Find $(4x - 7) - (2x - 2)$.

Find the GCF of each pair of monomials.

13. $3x$, $12x$

14. $16a$, $20ab$

15. $25cd$, $10d$

1. _____
2. _____
3. _____
4. _____
5. _____
6. _____
7. _____
8. _____
9. _____
10. _____
11. _____
12. _____
13. _____
14. _____
15. _____

Course 2 • Chapter 5 Expressions

NAME _____ DATE _____ PERIOD _____

Chapter Quiz

Evaluate each expression if $a = 3$, $b = 5$, and $c = 1$.

1. $b - c$

2. $ac + b$

3. $\dfrac{2(b + c)}{a}$

Use the Distributive Property to evaluate each expression.

4. $4(9 + 1)$

5. $-5(6 + x)$

Name the property shown by each statement.

6. $c \times 1 = c$

7. $83 + (52 + 17) = (83 + 52) + 17$

8. $22 + b + 18 = b + 22 + 18$

9. **JOGGING** Roberta jogs 3 laps the first day, 5 laps the second day, 7 laps the third day, and so on. On which day will Roberta jog 13 laps if the pattern continues?

Describe the relationship between the terms in each arithmetic sequence. Then write the next three terms in each sequence.

10. 2, 9, 16, 23, …

11. 50, 53, 56, 59, …

12. 0.2, 0.6, 1.0, 1.4, …

13. 81, 90, 99, 108, …

1. _____
2. _____
3. _____
4. _____
5. _____
6. _____
7. _____
8. _____
9. _____
10. _____
11. _____
12. _____
13. _____

Course 2 • Chapter 5 Expressions

NAME _____ DATE _____ PERIOD _____

Vocabulary Test

SCORE _____

Additive Identity Property	define a variable	Multiplicative Property of Zero
algebra	Distributive Property	property
algebraic expression	equivalent expressions	sequence
arithmetic sequence	factor	simplest form
Associative Property	factored form	simplify
coefficient	geometric sequence	term
Commutative Property	like terms	variable
constant	monomial	
counterexample	Multiplicative Identity Property	

Choose from the terms above to complete each sentence.

1. The numbers 2, 5, 8, 11, … are an example of a(n) _____.

    1. _____

2. The numerical factor of a multiplication expression that contains a variable is called a(n) _____.

    2. _____

3. Each number in a sequence is called a(n) _____.

    3. _____

4. A(n) _____ contains variables, numbers, and at least one operation.

    4. _____

5. Expressions that have the same value are called _____.

    5. _____

6. A(n) _____ is a term without a variable.

    6. _____

7. An algebraic expression is in _____ if it has no like terms and no parentheses.

    7. _____

8. _____ contain the same variables to the same powers.

    8. _____

Course 2 • Chapter 5 Expressions

103

Standardized Test Practice

Read each question. Then fill in the correct answer on the answer document provided by your teacher or on a sheet of paper.

1. What is $(3x - 2) - (4x + 1)$ in simplest form?

   A. $x - 3$
   B. $-x - 3$
   C. $-x + 1$
   D. $x + 1$

2. Roberto is training for the cross country team. The table shows the number of minutes he ran the first five days.

Day	Number of Minutes
Day 1	30
Day 2	30
Day 3	40
Day 4	40
Day 5	50

   If the pattern continues, which of the following shows the number of minutes he will run the next three days?
   F. 50, 50, 60
   G. 50, 60, 60
   H. 60, 60, 70
   I. 60, 70, 80

3. **GRIDDED RESPONSE** What is the value of the expression below if $x = 6$ and $y = 4$?

   $$(x + y) \div 5$$

4. Which of the following describes the relationship between the value of a term and n, its position in the sequence?

Position	1	2	3	4	5	n
Value of Term	3	6	9	12	15	■

   A. Add 2 to n.
   B. Divide n by 3.
   C. Multiply n by 3.
   D. Subtract n from 2.

5. **GRIDDED RESPONSE** Parker baked 80 cookies for a bake sale. At the sale, 70% of his cookies sold. How many of Parker's cookies were sold?

6. Which fraction is between $\frac{1}{2}$ and $\frac{3}{4}$?

   F. $\frac{1}{4}$ H. $\frac{3}{5}$
   G. $\frac{1}{3}$ I. $\frac{7}{8}$

7. What is the first step in evaluating the expression $3 \times (5 + 4) - 27 \div 9$?

   A. multiplying 3 and 5
   B. adding 5 and 4
   C. subtracting 27
   D. dividing 27 and 9

8. **GRIDDED RESPONSE** A square-shaped bulletin board is shown.

   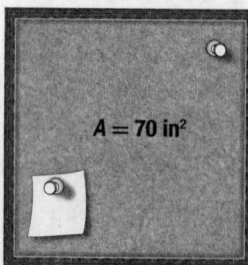

   $A = 70$ in²

   If a teacher covers 35% of the board with papers, how many square feet will not be covered?

9. What is the perimeter of the square garden?

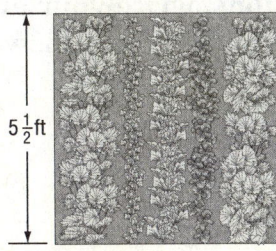
$5\frac{1}{2}$ ft

   F. 5 feet
   G. 20 feet
   H. 22 feet
   I. 30.25 feet

10. Sachi collects stamps. Each year, the number of stamps in her collection is ten times n, the number's position in the sequence. Which sequence represents Sachi's number of stamps?

   A. 1, 11, 21, 31
   B. 1, 10, 100, 1,000
   C. 10, 11, 12, 13
   D. 10, 20, 30, 40

11. What is the GCF of $45x^2y$ and $9x^3$?
   F. 9
   G. $9x$
   H. $9x^2$
   I. $3x^2$

12. **SHORT RESPONSE** Lemisha drove an average of 50 miles per hour on Sunday, 55 miles per hour on Monday, and 53 miles per hour on Tuesday. Let s represent the number of hours she drove on Sunday, m represent the number of hours she drove on Monday, and t represent the number of hours she drove on Tuesday. Write an expression that represents the total distance Lemisha drove.

13. Which of the following expressions can be written as $5(3 + x)$?
   A. $x \cdot 5 + x \cdot 3$
   B. $5 \cdot 3 + 5 \cdot x$
   C. $5 \cdot 3 + x$
   D. $3 + 5 \cdot x$

14. **SHORT RESPONSE** Use the Distributive Property to rewrite $4(12) + 4(8)$. Then evaluate the expression.

15. Which statement below is an example of the Associative Property of Addition?
   F. $7 + (3 + 5) = 7 + (5 + 3)$
   G. $9 + (11 + 6) = (9 + 11) + 6$
   H. $3(6 + 5) = 3 \cdot 6 + 3 \cdot 5$
   I. $12(8 + 4) = 12(8) + 12(4)$

16. **EXTENDED RESPONSE** The first and fifth terms of a sequence are shown.

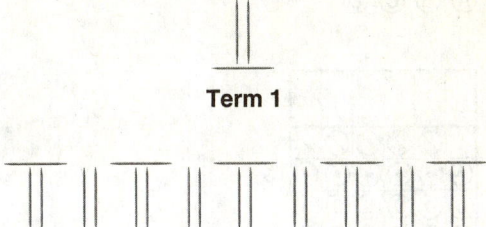
Term 1

Term 5

Part A What might the third term look like?

Part B Describe the relationship between the term number and the sequence.

Part C Write a rule that connects the term number and the number of toothpicks in the sequence.

Course 2 • Chapter 5 Expressions

NAME _____ DATE _____ PERIOD _____

Student Recording Sheet

SCORE _____

Use this recording sheet with the Standardized Test Practice pages.

Fill in the correct answer. For gridded-response questions, write your answers in the boxes on the answer grid and fill in the bubbles to match your answers.

1. Ⓐ Ⓑ Ⓒ Ⓓ

2. Ⓕ Ⓖ Ⓗ Ⓘ

3.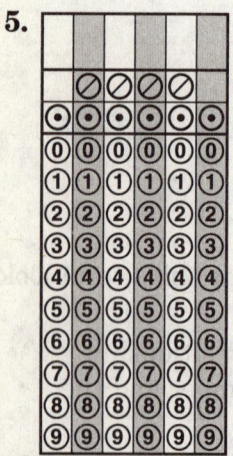

4. Ⓐ Ⓑ Ⓒ Ⓓ

5.

6. Ⓕ Ⓖ Ⓗ Ⓘ

7. Ⓐ Ⓑ Ⓒ Ⓓ

8.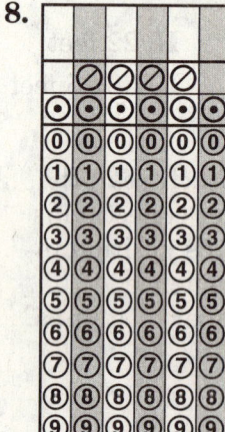

9. Ⓕ Ⓖ Ⓗ Ⓘ

10. Ⓐ Ⓑ Ⓒ Ⓓ

11. Ⓕ Ⓖ Ⓗ Ⓘ

12. _____

13. Ⓐ Ⓑ Ⓒ Ⓓ

14. _____

15. Ⓕ Ⓖ Ⓗ Ⓘ

Extended Response

Record your answers for Exercise 16 on the back of this paper.

106 Course 2 • Chapter 5 Expressions

NAME _____ DATE _____ PERIOD _____

Extended-Response Test

SCORE _____

Demonstrate your knowledge by giving a clear, concise solution to each problem. Be sure to include all relevant drawings and justify your answers. You may show your solution in more than one way or investigate beyond the requirements of the problem. If necessary, record your answer on another piece of paper.

1. Evaluate $7p + 6(p \div q) - 2q$ if $p = 6$ and $q = 3$. Show your work and give an explanation for each step.

2. **MOVIE TICKETS** For a class trip, a teacher bought 25 student tickets and 5 adult tickets.

   a. Write an expression for the total cost of the tickets purchased.

   b. If student tickets cost $4 each and adult tickets cost $6 each, how much money did the teacher spend on tickets?

3. Describe the pattern in the following sequence and explain how you would find the next term in the sequence.
   752, 756, 760, 764, 768, …

4. Explain in your own words what is meant by *combine like terms*.

Course 2 • Chapter 5 Expressions

107

NAME _____ DATE _____ PERIOD _____

Extended-Response Rubric

SCORE _____

Score	Description
4	A score of four is a response in which the student demonstrates a thorough understanding of the mathematics concepts and/or procedures embodied in the task. The student has responded correctly to the task, used mathematically sound procedures, and provided clear and complete explanations and interpretations. The response may contain minor flaws that do not detract from the demonstration of a thorough understanding.
3	A score of three is a response in which the student demonstrates an understanding of the mathematics concepts and/or procedures embodied in the task. The student's response to the task is essentially correct with the mathematical procedures used and the explanations and interpretations provided demonstrating an essential but less than thorough understanding. The response may contain minor flaws that reflect inattentive execution of mathematical procedures or indications of some misunderstanding of the underlying mathematics concepts and/or procedures.
2	A score of two indicates that the student has demonstrated only a partial understanding of the mathematics concepts and/or procedures embodied in the task. Although the student may have used the correct approach to obtaining a solution or may have provided a correct solution, the student's work lacks an essential understanding of the underlying mathematical concepts. The response contains errors related to misunderstanding important aspects of the task, misuse of mathematical procedures, or faulty interpretations of results.
1	A score of one indicates that the student has demonstrated a very limited understanding of the mathematics concepts and/or procedures embodied in the task. The student's response is incomplete and exhibits many flaws. Although the student's response has addressed some of the conditions of the task, the student reached an inadequate conclusion and/or provided reasoning that was faulty or incomplete. The response exhibits many flaws or may be incomplete.
0	A score of zero indicates that the student has provided no response at all, or a completely incorrect or uninterpretable response, or demonstrated insufficient understanding of the mathematics concepts and/or procedures embodied in the task. For example, a student may provide some work that is mathematically correct, but the work does not demonstrate even a rudimentary understanding of the primary focus of the task.

NAME _____ DATE _____ PERIOD _____

Test, Form 1A

SCORE _____

Write the letter for the correct answer in the blank at the right of each question.

1. What is the value of the expression $f^2 + 8$ if $f = 7$?
   A. 8   B. 15   C. 57   D. 78

   1. _____

2. What is the value of $6x - 3y$ if $x = 5$ and $y = -1$?
   F. 11   G. 33   H. 6^5 I. 65

   2. _____

3. Which property of multiplication is shown by the equation $f \cdot 8 = 8f$?
   A. Associative   C. Commutative
   B. Distributive   D. Identity

   3. _____

4. What are the next three terms in the sequence 4, 8, 12, 16, …?
   F. 18, 20, 21   H. 18, 22, 26
   G. 20, 24, 28   I. 32, 48, 64

   4. _____

5. What are the next three terms in the sequence 1, 7, 13, 19, …?
   A. 133, 931, 6517   C. 26, 33, 40
   B. 25, 31, 37   D. 19, 25, 31

   5. _____

6. What is the next term in the pattern 0.3, 0.5, 0.7, 0.9, …?
   F. 0.8   G. 0.1   H. 1.1   I. 2.1

   6. _____

7. What is the value of the expression $-5(3 + 4)$?
   A. −35   B. −12   C. −2   D. 2

   7. _____

Course 2 • Chapter 5 Expressions

NAME _____ DATE _____ PERIOD _____

Test, Form 1A (continued)   SCORE _____

8. Which expression is equivalent to $-5(x + 10)$?
   F. $5x + 50$ G. $-5x + 10$ H. $-5x - 50$ I. $5x - 50$ 8. _____

9. What is $-2y + 10 + 2y - 8$ simplified?
   A. 10   B. 8   C. 6   D. 2   9. _____

10. Simplify $x + 4 - 5x - 2$.
    F. $-5x + 2$ G. $-5x - 2$ H. $-4x + 2$ I. $-4x - 2$ 10. _____

11. Factor $15x + 35$ using the GCF.
    A. $5(3x + 7)$ B. $5(3x + 35)$ C. $15(x + 2)$ D. $7(2x + 5)$ 11. _____

12. Add $(11x + 2) + (9x - 4)$.
    F. $20x - 6$ G. $20x - 2$ H. $2x - 6$ I. $2x - 2$ 12. _____

13. Subtract $(8x + 6) - (x + 4)$.
    A. $7x + 2$ B. $7x + 10$ C. $9x + 10$ D. $9x + 2$ 13. _____

14. In the expression $12a + 5$, identify the coefficient.
    F. 12   G. a H. 5   I. 0   14. _____

110   Course 2 • Chapter 5 Expressions

NAME _____ DATE _____ PERIOD _____

Test, Form 1B

SCORE _____

Write the letter for the correct answer in the blank at the right of each question.

1. What is the value of the expression $f^2 + 5$ if $f = 3$?
   A. 8 **B.** 2 **C.** 6 **D.** 14

   1. _____

2. What is the value of $9x - 4y$ if $x = 3$ and $y = 2$?
   F. 12 **G.** 19 **H.** 9^3 **I.** 27

   2. _____

3. Which property of multiplication is shown by the equation $14 \cdot y \cdot 1 = 14y$?
   A. Associative **C.** Commutative
   B. Distributive **D.** Identity

   3. _____

4. What are the next three terms in the sequence 3, 6, 9, 12, ...?
   F. 24, 48, 96 **H.** 14, 16, 18
   G. 15, 18, 21 **I.** 15, 17, 20

   4. _____

5. What are the next three terms in the sequence 2.0, 3.5, 5.0, 6.5, ...?
   A. 8.5, 9.0, 11.5 **C.** 8.25, 11.75, 14.5
   B. 8.0, 9.5, 11.0 **D.** 7.0, 8.5, 9.0

   5. _____

6. What is the next term in the pattern 0.4, 0.8, 1.2, 1.6, ...?
   F. 1.7 **G.** 1.8 **H.** 1.9 **I.** 2.0

   6. _____

7. What is the value of the expression $3(2 - 7)$?
   A. −27 **B.** −15 **C.** 15 **D.** 27

   7. _____

Course 2 • Chapter 5 Expressions

NAME _____ DATE _____ PERIOD _____

Test, Form 1B (continued) SCORE _____

8. Which expression is equivalent to $-3(x + 10)$?
   F. $3x + 30$ G. $-3x + 10$ H. $-3x - 30$ I. $3x - 30$ 8. _____

9. What is $7y + 15 - 7y - 8$?
   A. 3   B. 5   C. 6   D. 7   9. _____

10. Simplify $x + 6 - 3x - 8$.
    F. $-2x - 2$ G. $-2x - 14$ H. $-3x - 14$ I. $-3x - 2$ 10. _____

11. Factor $25x + 45$ using the GCF.
    A. $x(25 + 45)$ B. $5(5x + 9)$ C. $25(x + 2)$ D. $5(5x + 45)$ 11. _____

12. Add $(12x + 5) + (3x - 7)$.
    F. $9x - 12$ G. $9x - 2$ H. $15x - 12$ I. $15x - 2$ 12. _____

13. Subtract $(3x + 5) - (x + 2)$.
    A. $2x + 7$ B. $2x + 3$ C. $-2x - 7$ D. $4x + 7$ 13. _____

14. In the expression $10a + 4$, identify the coefficient.
    F. 10   G. a H. 4   I. 0   14. _____

NAME _____ DATE _____ PERIOD _____

Test, Form 2A

SCORE _____

Write the letter for the correct answer in the blank at the right of each question.

1. Use the Distributive Property to write $2(-5 + 3)$ as an equivalent expression. What is the value of the expression?
   A. $2(8); 16$
   B. $2(-5) + 3; 13$
   C. $2(-5) + 2(3); -4$
   D. $(5 + 3)2; 16$

   1. _____

2. Add $(-x + 4) + (3x + 2)$.
   F. $2x + 6$ G. $2x + 2$ H. $8x + 6$ I. $4x + 6$

   2. _____

3. Subtract $(-x - 2) - (4x + 3)$.
   A. $3x - 1$ B. $-5x - 5$ C. $-5x - 2$ D. $3x - 2$

   3. _____

4. What is the value of $-4(9 + 6)$?
   F. 60 G. 30 H. -30 I. -60

   4. _____

5. What is the value of $9 + 2(5 - 3) - 6$?
   A. 1 B. 7 C. 9 D. 13

   5. _____

6. Simplify $4x + 8 + 2x - 7$.
   F. $6x + 15$ G. $6x - 1$ H. $6x + 1$ I. $2x + 1$

   6. _____

7. What is the value of $5a + 7b$ if $a = 4$ and $b = 3$?
   A. 41 B. 12 C. 43 D. 35

   7. _____

Course 2 • Chapter 5 Expressions

113

NAME _____ DATE _____ PERIOD _____

Test, Form 2A (continued)

SCORE _____

8. What are the next three terms in the sequence 15, 26, 37, 48, …?
   F. 57, 66, 75
   H. 58, 68, 78
   G. 58, 69, 70
   I. 59, 70, 81

   8. _____

9. What are the next three terms in the sequence 1.0, 2.5, 4.0, 5.5, …?
   A. 7.5, 8.0, 10.5
   C. 7.0, 8.5, 10.0
   B. 7.0, 9.5, 11.0
   D. 6.0, 7.5, 8.0

   9. _____

10. What is the value of $6d - 2c$ if $d = 3$ and $c = 4$?
    F. 10   G. 12   H. 14   I. 16

    10. _____

11. Evaluate $\frac{9r}{s}$ if $r = 6$ and $s = 3$.

    11. _____

12. Name the property of multiplication shown the equation $(3 \cdot x) \cdot 8 = 3 \cdot (x \cdot 8)$.

    12. _____

13. The area of a rectangular pool is $(15x - 9)$ square units. Factor $15x - 9$ to find possible dimensions of the pool.

    13. _____

14. Two friends bought a $14 pizza. Each person bought their own drink. The total cost of the food can be represented by the expression $2x + $14. What expression represents the cost of food for one person?

    14. _____

114   Course 2 • Chapter 5 Expressions

NAME _____ DATE _____ PERIOD _____

Test, Form 2B

SCORE _____

Write the letter for the correct answer in the blank at the right of each question.

1. Use the Distributive Property to write $-4(7 - 1)$ as an equivalent expression. What is the value of the expression?
   A. $4(6)$; 24
   B. $4(7) + 1$; 29
   C. $(7 + 1)4$; 32
   D. $-4(7) + -4(-1)$; -24

   1. _____

2. Add $(-2x + 4) + (4x + 6)$.
   F. $2x + 10$ G. $6x + 10$ H. $2x - 2$ I. $6x - 2$

   2. _____

3. Subtract $(-x - 3) - (5x + 2)$.
   A. $4x - 1$ B. $-6x - 5$ C. $4x - 5$ D. $-6x + 1$

   3. _____

4. What is the value of $-2(x + 5)$?
   F. $x + 10$ G. $x - 10$ H. $-2x - 2$ I. $-2x - 10$

   4. _____

5. What is the value of $11 + 3(9 - 6) - 4$?
   A. 12   B. 16   C. 18   D. 22

   5. _____

6. Simplify $2x + 1 - x + 2$.
   F. $3x + 3$ G. $x - 3$ H. $3x - 1$ I. $x + 3$

   6. _____

7. What is the value of $6a + 2b$ if $a = 1$ and $b = 2$?
   A. 7   B. 8   C. 10   D. 14

   7. _____

Course 2 • Chapter 5 Expressions

115

Test, Form 2B (continued)

8. What are the next three terms in the sequence 90, 84, 78, 72, ...?
   F. 68, 62, 56
   G. 66, 72, 78
   H. 66, 60, 54
   I. 66, 58, 52

   8. _____

9. What are the next three terms in the sequence 0.4, 0.9, 1.4, 1.9, ...?
   A. 2.4, 3.0, 3.6
   B. 2.2, 2.9, 3.2
   C. 2.2, 2.7, 3.2
   D. 2.4, 2.9, 3.4

   9. _____

10. What is the value of $12d - 7c$ if $d = 2$ and $c = 3$?
    F. 3   G. 7   H. 11   I. 13

    10. _____

11. Evaluate $\frac{4r}{s}$ if $r = 9$ and $s = 2$.

    11. _____

12. Name the multiplication property shown the equation $12 \cdot m = m \cdot 12$.

    12. _____

13. The area of a gymnasium is $(27x + 18)$ square units. Factor $27x + 18$ to find possible dimensions of the gymnasium.

    13. _____

14. Melanie spent four hours babysitting. She charges $16 upfront and then an additional charge per hour. The total amount she earned can be represented by the expression $4x + $16. What expression represents the amount she earns each hour?

    14. _____

NAME _____ DATE _____ PERIOD _____

Test, Form 3A

SCORE _____

Write the correct answer in the blank at the right of each question.

1. Name the property of multiplication that is shown by the equation $4x \cdot 1 = 4x$.

    1. _____

2. Use the Distributive Property to write $-2(n + 9)$ as an equivalent expression.

    2. _____

3. Name the property of addition that is shown by the equation $3 + b = b + 3$.

    3. _____

4. Simplify the expression $8(3x - 2) - 4(2x + 5)$.

    4. _____

5. Identify the next three terms in the arithmetic sequence 9, 13, 17, 21, …

    5. _____

6. Find the GCF of $14ab$, $7a$.

    6. _____

7. Add $(-12x + 6) + (4x - 12)$.

    7. _____

8. Subtact $(7x - 10) - (-6x - 8)$.

    8. _____

Course 2 • Chapter 5 Expressions

NAME _____ DATE _____ PERIOD _____

Test, Form 3A *(continued)* SCORE _____

9. Gloria bought each of her three younger siblings a movie ticket. Find the total cost of the movie tickets if each ticket cost $9.75. Justify your answer by using the Distributive Property.

9. _____

10. Write the expression $6a - 2(a - 1)$ in simplest form.

10. _____

11. Evaluate the expression $4(9 - 16)$.

11. _____

12. Evaluate the expression $7b - 5d$ if $b = 6$ and $d = 5$.

12. _____

13. Evaluate the expression $\frac{10b}{a}$ if $b = 6$ and $a = 2$.

13. _____

14. Marty bought a shirt for $9.50, sandals for $15.50, shorts for $11.75, and sunscreen for $5.25. Use mental math to find the total amount he spent. Explain.

14. _____

15. Explain why the expression $7x + 5$ cannot be factored.

15. _____

NAME _____ DATE _____ PERIOD _____

Test, Form 3B

SCORE _____

Write the correct answer in the blank at the right of each question.

1. Name the property of multiplication that is shown by the equation $17x \cdot 1 = 17x$.

   1. _____

2. Use the Distributive Property to write $6(w + 5)$ as an equivalent expression.

   2. _____

3. Name the property that is shown by the equation $5(4 + g) = 5(4) + 5(g)$.

   3. _____

4. Simplify the expression $-3(4x + 5) - 2(-5x + 8)$

   4. _____

5. Identify the next three terms in the arithmetic sequence 4.2, 4.8, 5.4, 6.0, …

   5. _____

6. Find the GCF of $6ab$, $3a$.

   6. _____

7. Add $(-7x + 1) + (2x - 11)$.

   7. _____

8. Subtract $(11x - 9) - (-4x - 6)$.

   8. _____

Course 2 • Chapter 5 Expressions

119

Test, Form 3B (continued)

9. Caroline bought 5 bracelets. Find the total cost of the bracelets if each bracelet cost $3.10. Justify your answer by using the Distributive Property.

9. _____

10. Write the expression $7b - 3(b - 1)$ in simplest form.

10. _____

11. Evaluate the expression $3(9 - 16)$.

11. _____

12. Evaluate the expression $4b - 6d$ if $b = 7$ and $d = 3$.

12. _____

13. Evaluate the expression $\frac{8b}{a}$ if $b = 8$ and $a = 2$.

13. _____

14. Justin sent 24 text messages on Monday, 48 on Tuesday, 36 on Wednesday, and 22 on Thursday. Use mental math to find the total number of messages he sent during the four days. Explain.

14. _____

15. Explain why the expression $13x + 10$ cannot be factored.

15. _____

Are You Ready?

Review

Words and phrases in problems often suggest addition, subtraction, multiplication, and division.

Addition	Subtraction	Multiplication	Division
sum	minus	times	divided by
plus	less than	product	half
more than	decreased by	twice	quotient
increased by	difference	each	divide
in all	less	multiplied	per
			separate

Example 1

Write an algebraic expression for *6 more than a number*.

More than means add. Let n represent the number.

$6 + n$

Example 2

Write an algebraic expression for *4 less than the number of runs scored*.

Less than means subtract. Let r represent the number of runs.

$r - 4$

Exercises

Write each phrase as an algebraic expression.

1. the number of minutes decreased by 10

2. 12 pounds more than the weight of a watermelon

3. half the number of golf balls

4. the number of blocks divided by 4

5. twice the number of stuffed animals

6. 8 less than the number of songs

7. the number of miles increased by 50

8. 7 times the number of cats

Course 2 • Chapter 6 Equations and Inequalities

Are You Ready?

Practice

Write each phrase as an algebraic expression.

1. 15 more than a number

2. 9 less than the number of starfish

3. half the number of pencils

4. the number of chairs increased by 7

Identify the solution of each equation from the list given.

5. $m - 10 = 40$; 30, 40, 50

6. $2 \times s = 12$; 4, 6, 8

7. $f + 6 = 18$; 12, 13, 14

8. $n - 20 = 8$; 26, 28, 30

Solve mentally.

9. **DISTANCE** The equation $230 = d - 240$ describes the distance in miles between Pensacola and Tampa. If d is the distance between the two cities, how many miles apart are Pensacola and Tampa?

10. **VISITORS** The number of visitors at the aquarium was 3 times the number of visitors at the art museum. There were 1,500 visitors at the aquarium. Solve the equation $3m = 1,500$ to find the number of visitors m at the art museum.

Are You Ready?

Apply

1. **STADIUMS** There were 75,000 people in *The Swamp* one Saturday afternoon. The number of the people in *The Swamp* was 1.5 times that of the people in Dolphin Stadium. Solve the equation $1.5p = 75,000$ to find the number of people p in Dolphin Stadium.

2. **ALLIGATOR** Linda saw an alligator in the zoo. Dylon told her he had seen an alligator 3 feet longer than the one she saw. The alligator Dylon saw was 12 feet long. Solve the equation $\ell + 3 = 12$ to find the length ℓ of the alligator Linda saw.

3. **FLYING DISK** Marisha threw a flying disk 75 feet. This was 10 feet less than the distance that Quame threw a disk. Solve the equation $d - 10 = 75$ to find the distance d Quame threw the disk.

4. **VACATION** Twice as many days as the Jacobi were on vacation was 22. Solve the equation $2v = 22$ to find the number of days v they were on vacation.

5. **ENROLLMENT** The number of students in the high school is 300 more than the number of students at the junior high school. There are 900 students at the high school. Solve the equation $900 = j + 300$ to find the number j of students at the junior high.

6. **WALKING** Tenisha walked 5 blocks to school. The number of blocks Zeke walked to school decreased by 3 was equal to the number of blocks Tenisha walked to school. Solve the equation $z - 3 = 5$ to find the number of blocks z Zeke walked to school.

NAME _____ DATE _____ PERIOD _____

Diagnostic Test

Write each phrase as an algebraic expression.

1. 14 more points than the Dolphins scored

    1. _____

2. $3 more than Sara has

    2. _____

3. a number decreased by 12

    3. _____

Identify the solution of each equation from the list given.

4. $6 + n = 20$; 14, 15, 16

    4. _____

5. $p - 2 = 19$; 17, 19, 21

    5. _____

6. $4h = 24$; 6, 8, 9

    6. _____

7. $3s = 36$; 10, 11, 12

    7. _____

8. $c - 47 = 30$; 74, 77, 80

    8. _____

Solve mentally.

9. **ELEVATION** The highest elevation in Florida is 105 meters. The highest elevation in Georgia is about 5,000 meters higher than that in Florida. Solve the equation $g - 5,000 = 105$ to find the highest elevation g in Georgia.

    9. _____

10. **DRIVEWAY** The Grzelak's driveway is twice as long as the Lubeck's driveway. The Grzelak's driveway is 300 feet long. Solve the equation $2d = 300$ to find the length d of the Lubeck's driveway.

    10. _____

NAME _____ DATE _____ PERIOD _____

Pretest

Solve each equation. Check your solution.

1. $x + 7 = 12$

2. $x + 9 = 2$

3. $x - 4 = 5$

4. $36 = 9x$

5. $-3a = 15$

6. $\frac{b}{-3} = -7$

7. $12 = 0.3n$

8. $\frac{2}{3}x = \frac{8}{15}$

9. $4x + 6 = 10$

10. $3x - 10 = 5x + 20$

1. _____
2. _____
3. _____
4. _____
5. _____
6. _____
7. _____
8. _____
9. _____
10. _____

Find the multiplicative inverse, or reciprocal, of each number.

11. $\frac{3}{7}$

12. $3\frac{1}{4}$

11. _____
12. _____

Write an inequality and solve each problem.

13. Five more than a number is at most eleven.

14. The product of a number and 2 is at least twenty.

13. _____
14. _____

Solve each inequality. Graph the solution on a number line.

15. $x - 3 \leq 9$

15. _____
    8 9 10 11 12 13 14 15 16

16. $-5m > 20$

16. _____
    $-8\ -7\ -6\ -5\ -4\ -3\ -2\ -1\ 0$

Course 2 • Chapter 6 Equations and Inequalities

NAME _____ DATE _____ PERIOD _____

Chapter Quiz

Solve each equation. Check your solution.

1. $t + 7 = 5$

2. $a - 3 = 9$

3. $-4 = p - 6$

4. $y + 19 = 26$

5. $3x = -12$

6. $20 = 4n$

7. $2.1d = 51.24$

8. $\frac{3}{4}x = -12$

9. $\frac{5}{2}c = 8\frac{1}{3}$

10. $3y + 8 = 14$

11. $2m - 5 = 17$

12. $3x - 2.6 = x + 8.2$

13. **PLUMBER** Abe, a plumber, charges $50 per hour plus $75 for making an in-home visit. The Plumbing Service charges $65 per hour but no set fee for a visit. How many hours of plumbing work are needed for each to cost the same?

14. **BROWNIES** Lauren brings brownies to class one day. She gives $\frac{3}{5}$ of them to her friends Gareth, Tammy, and Camilla. If she gave them 21 brownies, how many did she bring to class?

1. _____
2. _____
3. _____
4. _____
5. _____
6. _____
7. _____
8. _____
9. _____
10. _____
11. _____
12. _____

13. _____

14. _____

NAME _____ DATE _____ PERIOD _____

Vocabulary Test

SCORE _____

Addition Property of Equality	inequality
Addition Property of Inequality	Multiplication Property of Equality
coefficient	Multiplication Property of Inequality
Division Property of Equality	solution
Division Property of Inequality	Subtraction Property of Equality
equation	Subtraction Property of Inequality
equivalent equation	two-step equation
formula	

Write a term or number to make each a true sentence.

1. A(n) _____ is a statement that two quantities are equal.   1. _____

2. The statements $3x + 2 = 17$ and $x = 5$ are _____.   2. _____

3. A(n) _____ is a statement that two quantities are not equal.   3. _____

4. The _____ states when you add the same number to each side of an inequality, the inequality stays true.   4. _____

5. A(n) _____ is an equation that has two operations.   5. _____

6. The _____ of x in the expression $8x$ is 8.   6. _____

7. The _____ states when you subtract the same number from each side of an equation, the equation remains equal.   7. _____

8. A(n) _____ is an equation that shows the relationship among certain quantities.   8. _____

Define each term in your own words.

9. Multiplication Property of Inequality   9. _____

10. Multiplication Property of Equality   10. _____

Course 2 • Chapter 6 Equations and Inequalities

Standardized Test Practice

Read each question. Then fill in the correct answer on the answer sheet provided by your teacher or on a sheet of paper.

1. A sports store sells two different field hockey kits shown in the table.

Hockey Kits	
Beginner	Basic
hockey stick	hockey stick
ball	ball
shin guards	

   The beginner's field hockey kit costs $150. It is $15 more than three times the cost of the basic kit. What is the cost of the basic kit?

   A. $35.00 **C.** $45.00
   B. $40.00 **D.** $50.00

2. Which line contains the ordered pair $(-2, 4)$?

   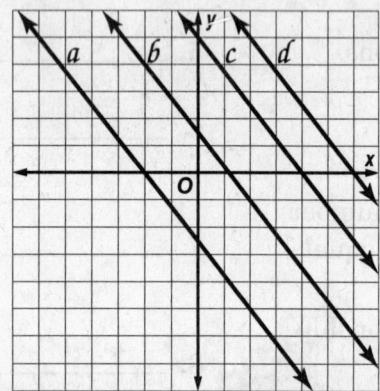

   F. line a **H.** line b
   G. line c **I.** line d

3. Which integer added to 12 gives a sum of -14?

   A. -18 **C.** -24
   B. -20 **D.** -26

4. **GRIDDED RESPONSE** Mrs. McDowell is making a big batch of cookies for her son's birthday. The price of the chocolate chips is 2 bags for $4.00. Use the table to determine the number of bags of chocolate chips r that Mrs. McDowell bought if the cost c was $12.

r	$r(4 \div 2)$	c
1	$1(4 \div 2)$	$2
2	$2(4 \div 2)$	$4
3	$3(4 \div 2)$	$6

5. **GRIDDED RESPONSE** Aida bought a costume box containing 50 costumes for $300. She sold all of the costumes and made a $250 profit. She sold all of the costumes for the same price. Use the equation $50c - 300 = 250$, where c is the selling price of each costume. What was the selling price of one costume in dollars?

6. Which of the following problems can be solved using the equation $x - 9 = 15$?

   F. Allison is 9 years younger than her sister Pam. Allison is 15 years old. What is x, Pam's age?

   G. David's portion of the bill is $9 more than Jaleel's portion of the bill. If Jaleel pays $9, find x, the amount in dollars that David pays.

   H. The sum of two numbers is 15. If one of the numbers is 9, what is x, the other number?

   I. Calvin owns 15 CDs. If he gave 9 of them to a friend, what is x, the number of CDs he has left?

7. What value of x makes this equation true?
$$4x + 7 = 43$$
   A. 12
   B. 10
   C. 9
   D. 8

8. Joshua spends $0.25 for every song he downloads to his cell phone. Which of the following represents the number of songs he can download if he has at least $3?

   F.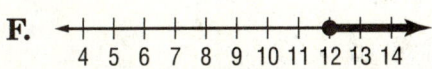
   G. (number line 4–14, open circle at 12, shaded right)
   H. (number line 4–14, open circle at 11, shaded left)
   I. Not enough information is given.

9. **SHORT RESPONSE** Rico, Carolina, and Gloria have pizza that they are going to be sharing with other people. Rico gave away $\frac{1}{3}$ of his cheese pizza to Carolina and she gave him $\frac{3}{7}$ of her pepperoni. Rico then gave Gloria $\frac{1}{7}$ of his cheese pizza. How much pizza, pepperoni and cheese, does Rico have now?

10. For a warm up, Samuel runs 200 yards less than half the maximum distance he can run. This is represented by the equation $r = \frac{1}{2}x - 200$, where x represents the maximum distance he can run and r represents the distance run during his warm up. If Samuel ran 1,600 yards during his warm up, what is the maximum distance he can run?
    A. 3,600 yards
    B. 2,400 yards
    C. 1,800 yards
    D. 1,600 yards

11. What is the value of $20 \div (-4)$?
    F. -5 H. 5
    G. -7 I. 4

12. **GRIDDED RESPONSE** Ines is in a hot air balloon 89 feet above the ground. A bird is flying 15 feet above the hot air balloon. How high off the ground is the bird in feet?

13. **EXTENDED RESPONSE** A first-time bungee jumper is about to make his first jump. When the bungee jumper jumps, he will fall 5 feet every 0.5 second.

    Part A Let s be the total number of seconds in a jump and h be the height of the jump. Write an equation that can be used to find s.

    Part B Use your equation to calculate the total seconds for a 150-foot jump. Show your work.

Course 2 · Chapter 6 Equations and Inequalities

NAME _____ DATE _____ PERIOD _____

Student Recording Sheet

SCORE _____

Use this recording sheet with the Standardized Test Practice pages.

Fill in the correct answer. For Gridded-Response questions, write your answers in the boxes on the answer grid and fill in the bubbles to match your answers.

1. Ⓐ Ⓑ Ⓒ Ⓓ
2. Ⓕ Ⓖ Ⓗ Ⓘ
3. Ⓐ Ⓑ Ⓒ Ⓓ

4.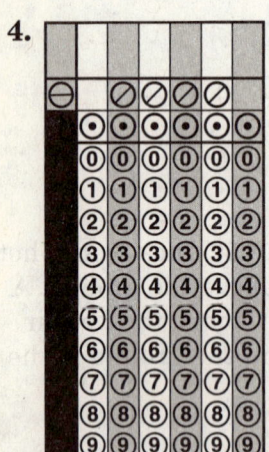

5.

6. Ⓕ Ⓖ Ⓗ Ⓘ
7. Ⓐ Ⓑ Ⓒ Ⓓ
8. Ⓕ Ⓖ Ⓗ Ⓘ
9. _____
10. Ⓐ Ⓑ Ⓒ Ⓓ
11. Ⓕ Ⓖ Ⓗ Ⓘ

12.

Extended Response
Record your answers for Exercise 13 on the back of this paper.

130   Course 2 • Chapter 6 Equations and Inequalities

NAME _____ DATE _____ PERIOD _____

SCORE _____

Extended-Response Test

Demonstrate your knowledge by giving a clear, concise solution to each problem. Be sure to include all relevant drawings and justify your answers. You may show your solution in more than one way or investigate beyond the requirements of the problem. If necessary, record your answer on another piece of paper.

1. The Ortiz family will travel round-trip from Indianapolis to Denver this summer. The driving distance between the two cities is 1,058 miles. They need to find out about how much the gas will cost round trip. Their car gets about 20 miles per gallon. They usually spend about $2.89 per gallon.

    a. Write and solve an algebraic equation to find out how many miles they will travel. Explain your steps.

    b. Write and solve an algebraic equation to find out how many gallons of gas they will use. Explain your steps.

    c. Write and solve an algebraic equation to find out how much the Ortiz's will spend on gas round trip. Explain your steps.

    d. Suppose you know the total cost of the gas, but do not know the price per gallon. Explain how you would work backward to solve the problem.

2. The students at Camden Middle School went on a class ski trip. Each student paid $125 and the school paid a total of $300. Altogether, the trip cost at most $2,425. How many students went on the trip?

    a. Write an inequality for this situation.

    b. Solve the inequality and explain what the solution means.

Course 2 • Chapter 6 Equations and Inequalities

NAME _____ DATE _____ PERIOD _____

Extended-Response Rubric

SCORE _____

Score	Description
4	A score of four is a response in which the student demonstrates a thorough understanding of the mathematics concepts and/or procedures embodied in the task. The student has responded correctly to the task, used mathematically sound procedures, and provided clear and complete explanations and interpretations. The response may contain minor flaws that do not detract from the demonstration of a thorough understanding.
3	A score of three is a response in which the student demonstrates an understanding of the mathematics concepts and/or procedures embodied in the task. The student's response to the task is essentially correct with the mathematical procedures used and the explanations and interpretations provided demonstrating an essential but less than thorough understanding. The response may contain minor flaws that reflect inattentive execution of mathematical procedures or indications of some misunderstanding of the underlying mathematics concepts and/or procedures.
2	A score of two indicates that the student has demonstrated only a partial understanding of the mathematics concepts and/or procedures embodied in the task. Although the student may have used the correct approach to obtaining a solution or may have provided a correct solution, the student's work lacks an essential understanding of the underlying mathematical concepts. The response contains errors related to misunderstanding important aspects of the task, misuse of mathematical procedures, or faulty interpretations of results.
1	A score of one indicates that the student has demonstrated a very limited understanding of the mathematics concepts and/or procedures embodied in the task. The student's response is incomplete and exhibits many flaws. Although the student's response has addressed some of the conditions of the task, the student reached an inadequate conclusion and/or provided reasoning that was faulty or incomplete. The response exhibits many flaws or may be incomplete.
0	A score of zero indicates that the student has provided no response at all, or a completely incorrect or uninterpretable response, or demonstrated insufficient understanding of the mathematics concepts and/or procedures embodied in the task. For example, a student may provide some work that is mathematically correct, but the work does not demonstrate even a rudimentary understanding of the primary focus of the task.

NAME _____ DATE _____ PERIOD _____

Test, Form 1A

SCORE _____

Write the letter for the correct answer in the blank at the right of each question.

Solve.

1. Three times Geoff's age plus 3 is Myrka's age. Myrka is 48. What is Geoff's age?
   A. 15   B. 17   C. 135   D. 153

   1. _____

2. In the basketball game, Rachael scored 6 points less than twice the number of points Trina scored. Trina scored 9 points. How many points did Rachael score?
   F. 3 points   G. 12 points   H. 15 points   I. 18 points

   2. _____

What is the solution of each equation?

3. $9 + n = -2$
   A. -11 B. -7 C. 2   D. 7

   3. _____

4. $14 = y - 10$
   F. -24 G. -4 H. 4   I. 24

   4. _____

5. $5 = x + 3$
   A. -8 B. -2 C. 2   D. 8

   5. _____

6. $t - 26 = -21$
   F. -47 G. -5 H. 5   I. 47

   6. _____

7. $84 = 7d$
   A. 8   B. 12   C. 77   D. 91

   7. _____

8. $6z = 12$
   F. 2   G. 6   H. 18   I. 72

   8. _____

9. $\frac{x}{4} = -1$
   A. -4 B. $-\frac{1}{4}$ C. $\frac{1}{4}$ D. 4

   9. _____

10. $-2 + \frac{4}{5}x = 6$
    F. -5 G. 5   H. 8   I. 10

    10. _____

11. $\frac{2}{3}x = 4$
    A. 6   B. $3\frac{1}{3}$ C. $2\frac{2}{3}$ D. $1\frac{1}{2}$

    11. _____

Course 2 • Chapter 6 Equations and Inequalities

Test, Form 1A (continued)

12. $\frac{1}{2}x = \frac{5}{6}$
    F. $\frac{1}{3}$
    G. $\frac{5}{12}$
    H. $1\frac{1}{6}$
    I. $1\frac{2}{3}$

    12. _____

13. $-8x + 3 = -29$
    A. 256
    B. 4
    C. 3
    D. -40

    13. _____

14. $3x + 1 = -11$
    F. -36
    G. -30
    H. -4
    I. -3

    14. _____

15. $\frac{5}{6}(x + 4) = 15$
    A. -18
    B. -14
    C. 14
    D. 18

    15. _____

16. $0.3a = 6$
    F. 1.8
    G. 2
    H. 18
    I. 20

    16. _____

17. What is the solution of $x + 6 < 5$?
    A. $x < 11$
    B. $x > 11$
    C. $x < -1$
    D. $x > -1$

    17. _____

18. What is the solution of $\frac{y}{-2} \leq 3$?
    F. $y \leq -6$
    G. $y \geq -6$
    H. $y \leq -\frac{3}{2}$
    I. $y \geq -\frac{3}{2}$

    18. _____

19. What is the solution of $\frac{2}{3}a + 6 > 0$?
    A. $a > -9$
    B. $a < -9$
    C. $a > -4$
    D. $a < -4$

    19. _____

20. What is the graph of the solution set of $b - 4 \geq -1$?

    F. number line from -3 to 5, closed circle at 3, shaded right

    G. number line from -3 to 5, closed circle at 3, shaded left

    H. number line from -6 to 3, open circle at -5, shaded right

    I. number line from -6 to 3, open circle at -5, shaded right

    20. _____

NAME _____ DATE _____ PERIOD _____

SCORE _____

Test, Form 1B

Write the letter for the correct answer in the blank at the right of each question.

Solve.

1. Four times Audrey's age plus 7 is Parag's age. Parag is 51. What is Audrey's age?
   A. 7   B. 11   C. 15   D. 212

   1. _____

2. Larue scored 10 points less than 3 times the number of points that Ross scored. Larue scored 11 points. How many points did Ross score?
   F. 7   G. 19   H. 21   I. 23

   2. _____

What is the solution of each equation?

3. $7 + n = -5$
   A. -12 B. -2 C. 2   D. 12

   3. _____

4. $11 = y - 6$
   F. -17 G. -5 H. 5   I. 17

   4. _____

5. $8 = x + 1$
   A. 9   B. 7   C. -7 D. -9

   5. _____

6. $t - 34 = -15$
   F. 49   G. 19   H. -19 I. -49

   6. _____

7. $96 = 12d$
   A. 108   B. 84   C. 8   D. $\frac{1}{8}$

   7. _____

8. $4z = 20$
   F. 16   G. 5   H. 4   I. -16

   8. _____

9. $\frac{y}{5} = -2$
   A. 10   B. 3   C. $-\frac{2}{5}$ D. -10

   9. _____

10. $7 + \frac{7}{9}x = -42$
    F. -63 G. -49 H. 49   I. 63

    10. _____

11. $\frac{3}{5}x = 18$
    A. $10\frac{4}{5}$ B. $17\frac{2}{5}$ C. 30   D. 90

    11. _____

Course 2 • Chapter 6 Equations and Inequalities

135

Test, Form 1B (continued)

12. $\frac{1}{3}a = \frac{7}{9}$
    F. $2\frac{1}{3}$ G. $1\frac{2}{7}$ H. $\frac{7}{27}$ I. $\frac{1}{3}$ 12. _____

13. $4x + 1 = -15$
    A. $-3\frac{1}{2}$ B. -4 C. -12 D. -20 13. _____

14. $6x + 9 = 7x - 5$
    F. 14   G. 4   H. -4 I. -14 14. _____

15. $-0.4(x + 3) = 4$
    A. -13 B. -0.6 C. 1.4 D. 7 15. _____

16. $0.4p = -9$
    F. -36 G. -22.5 H. -9.4 I. -8.6 16. _____

17. What is the solution of $x + 9 < 5$?
    A. $x < 4$ B. $x > 4$ C. $x < -4$ D. $x > -4$ 17. _____

18. What is the solution of $\frac{y}{-2} \leq 5$?
    F. $y \leq \frac{-5}{2}$ G. $y \geq \frac{-5}{2}$ H. $y \leq -10$ I. $y \geq -10$ 18. _____

19. What is the solution of $\frac{2}{3}a + 4 > -8$?
    A. $a > -8$ B. $a < -8$ C. $a > -18$ D. $a < -18$ 19. _____

20. What is the graph of the solution set of $a - 5 \leq -1$?

    F. [number line from -7 to 3, closed circle at -6, shaded right]

    G. [number line from -7 to 3, closed circle at -6, shaded left]

    H. [number line from -5 to 5, open circle at 4, shaded right]

    I. [number line from -5 to 5, closed circle at 4, shaded left]

    20. _____

NAME _____ DATE _____ PERIOD _____

Test, Form 2A

SCORE _____

Write the letter for the correct answer in the blank at the right of each question.

1. Edwin's mother is 57 years old. Her age is three years more than twice Edwin's age. What is Edwin's age?
   A. 30 years   B. 27 years   C. 15 years   D. 37 years

   1. _____

2. In a basketball game, Benito scored 3 points less than twice the number of points Carnell scored. Carnell scored 8 points. How many points did Benito score?
   F. 5 points   G. 11 points   H. 13 points   I. 16 points

   2. _____

What is the solution of each equation?

3. $t + 16 = 7$
   A. −23   B. −9   C. 9   D. 23

   3. _____

4. $\frac{w}{4} = -11$
   F. −44   G. −7   H. 7   I. 44

   4. _____

5. $81 = 3k$
   A. 27   B. 78   C. 84   D. 243

   5. _____

6. $-8 + \frac{5}{6}x = -28$
   F. −24   G. −20   H. −4   I. 24

   6. _____

7. $\frac{1}{2}z = 9\frac{1}{4}$
   A. $4\frac{5}{8}$ B. $9\frac{1}{8}$ C. $18\frac{1}{4}$ D. $18\frac{1}{2}$

   7. _____

8. $37 = 18q + 1$
   F. 0.5   G. 2   H. 12   I. 19

   8. _____

9. $2y - 1.7 = 3.3$
   A. 0.8   B. 2.5   C. 3.2   D. 10

   9. _____

10. The length of each side of a square was decreased by 2 inches, so the perimeter is now 48 inches. What was the original length of each side of the square?
    F. 10 in.   G. 12 in.   H. 14 in.   I. 16 in.

    10. _____

Course 2 • Chapter 6 Equations and Inequalities

NAME _____ DATE _____ PERIOD _____

Test, Form 2A (continued)

SCORE _____

What is an equivalent equation for each given equation?

11. $4x + 11 = 15$
    - **A.** $4x = 26$
    - **B.** $-4x + 11 = -15$
    - **C.** $15 = -11 - 4x$
    - **D.** $4x = 4$

11. _____

12. $3(x - 9) = 3$
    - **F.** $3x - 9 = 3$
    - **G.** $3x - 9 = 9$
    - **H.** $x - 9 = 1$
    - **I.** $x - 9 = 9$

12. _____

Solve each equation. Check your solution.

13. $0.3a = 51$

13. _____

14. $\dfrac{z}{5.2} = 16$

14. _____

15. $-11 = x + 5$

15. _____

16. $3j = 2.7$

16. _____

17. $\dfrac{x}{8} - 2 = -3$

17. _____

18. $-0.4(x - 3.8) = -2$

18. _____

19. Each deli sandwich made uses $\frac{1}{4}$ pound of turkey. Tyler started with $3\frac{7}{8}$ pounds of turkey and now has $1\frac{5}{8}$ pounds left. How many deli sandwiches did he make?

19. _____

Solve each inequality. Graph the solution on a number line.

20. $x + 7 \leq 11$

20. _____

21. $m - 4 > 12$

21. _____

22. $3p \geq -24$

22. _____

23. $-2w + 5 < -5$

23. _____

24. $\dfrac{h}{-6} > 3$

24. _____

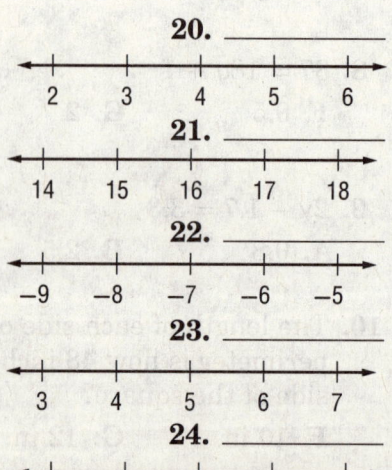

138 Course 2 • Chapter 6 Equations and Inequalities

NAME _____ DATE _____ PERIOD _____

Test, Form 2B

SCORE _____

Write the letter for the correct answer in the blank at the right of each question.

1. The Leungs sold a valuable painting for $55,000. This price is $1,000 more than twice the amount they originally paid for it. How much did they originally pay?
    A. $25,000   B. $27,000   C. $27,500   D. $28,000

    1. _____

2. In a volleyball game, Alexis scored 4 points more than twice the number of points Jessica scored. Jessica scored 3 points. How many points did Alexis score?
    F. 1 point   G. 7 points   H. 10 points   I. 12 points

    2. _____

What is the solution of each equation?

3. $m - 12 = 11$
    A. -23 B. -1 C. 1 D. 23

    3. _____

4. $\dfrac{x}{-5} = -6$
    F. -30 G. -11 H. 11 I. 30

    4. _____

5. $6x = -48$
    A. -8 B. -7 C. 7 D. 8

    5. _____

6. $7 + \dfrac{2}{5}x = 1$
    F. -15 G. -6 H. -3 I. 15

    6. _____

7. $\dfrac{1}{3}y = 4\dfrac{5}{6}$
    A. $1\dfrac{11}{8}$ B. $4\dfrac{5}{18}$ C. $12\dfrac{5}{6}$ D. $14\dfrac{1}{2}$

    7. _____

8. $-12 = 4.7k + 11.5$
    F. -5 G. -3 H. -0.1 I. 5

    8. _____

9. $-3m - 21 = -6$
    A. -45 B. -5 C. 9 D. -81

    9. _____

10. The length of each side of a square was increased by 6 inches, so the perimeter is now 52 inches. What was the original length of each side of the square?
    F. 1 in.   G. 7 in.   H. 13 in.   I. 19 in.

    10. _____

Course 2 • Chapter 6 Equations and Inequalities

Test, Form 2B (continued)

What is an equivalent equation for the given equation?

11. $6x + 10 = 23$
    - A. $6x = 13$
    - B. $6x = 33$
    - C. $23 = 10 - 6x$
    - D. $-23 = -6x + 10$

    11. _____

12. $-4(x - 11) = 16$
    - F. $x - 11 = 4$
    - G. $x - 11 = -4$
    - H. $-4x - 11 = 16$
    - I. $-4x + 44 = 64$

    12. _____

Solve each equation. Check your solution.

13. $0.2a = 48$

    13. _____

14. $\dfrac{z}{8.9} = 22$

    14. _____

15. $y + 9 = -21$

    15. _____

16. $2.7a = 13.5$

    16. _____

17. $\dfrac{b}{7} - 4 = -5$

    17. _____

18. $\dfrac{7}{10}(x - 4) = 42$

    18. _____

19. Each deli sandwich made uses $\dfrac{3}{8}$ pound of roast beef. Marisa started with 5 pounds of roast beef and now has $2\dfrac{3}{4}$ pounds left. How many deli sandwiches did she make?

    19. _____

Solve each inequality. Graph the solution on a number line.

20. $x + 5 \leq 13$

21. $m - 1 > 8$

22. $4p \geq -28$

23. $-3w - 2 < -17$

24. $\dfrac{h}{-2} > 6$

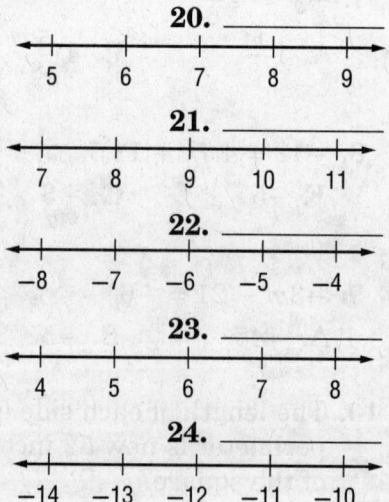

20. _____

21. _____

22. _____

23. _____

24. _____

140   Course 2 • Chapter 6 Equations and Inequalities

NAME _____ DATE _____ PERIOD _____

Test, Form 3A

SCORE _____

Solve each equation. Check your solution.

1. $x + 17.3 = -4.7$

2. $21.8 - g = 9.5$

Solve.

3. The number of infielders on a baseball team is one less than three times the number of pitchers. If there are eleven infielders, how many pitchers are there?

4. Three children each had the same amount of money in their savings accounts. One of the children withdrew a quarter of her money and spent it all on a $25 T-shirt. What was the total amount of money originally in the accounts?

Solve each equation. Check your solution.

5. $2.9a = 11.6$

6. $-3.1u = 7.75$

7. $-3.7x = 29.6$

8. $-\frac{4}{9}x = \frac{16}{27}$

9. $(x - 5)(-3) = 18$

10. $6m + 2.3 = -9.7$

11. $3.6 = -2p + 5.8$

12. $\frac{x}{4} - 7 = -2$

13. $2x + 6 = -4x$

14. $x - 17 = 2x + 3$

15. $21 - 6x = -11 - 14x$

Course 2 • Chapter 6 Equations and Inequalities

Test, Form 3A (continued)

Express an equivalent equation for each equation.

16. $-3x + 9 = 24$

16. _____

17. $\frac{5}{12}(x - 5) = 30$

17. _____

Solve each inequality. Graph the solution on a number line.

18. $y + 6 > 20$

18. _____
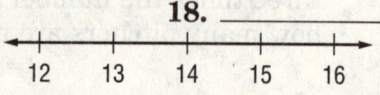

19. $w - 11 \leq 3$

19. _____
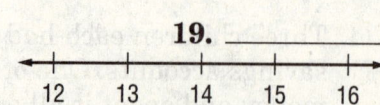

20. $5k \geq -45$

20. _____
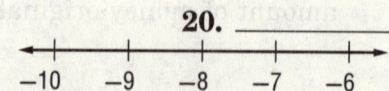

21. $-2q < -34$

21. _____
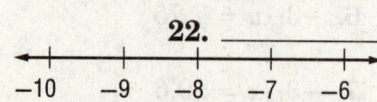

22. $\frac{p}{-1} > 7$

22. _____
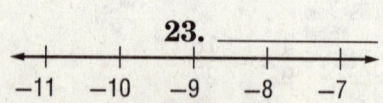

23. $\frac{b}{3} + 5 \leq 2$

23. _____

Write and solve an equation for Exercise 24.

24. The length of each side of an equilateral triangle is decreased by 4 inches, so the perimeter is now 27 inches. What is the original length of each side of the equilateral triangle?

24. _____

25. There are 256 vehicles in a car dealership's lot. At least 113 of them are hybrid vehicles. Write and solve an inequality that describes how many vehicles, at most, are not hybrid.

25. _____

26. It costs Guido $0.20 to send a text message from his cell phone. He has already spent $4 in text messages this month. If he has a total of $10 that he can spend this month on text messages, write and solve an inequality that will give the greatest number of text messages that he can send. Interpret the solution.

26. _____

Test, Form 3B

Solve each equation. Check your solution.

1. $m - 16 = -4$

2. $k + 31 = 17$

Solve.

3. Jamar earned $2,500 from his summer job at the grocery store. This is $350 more than twice what his friend Todd earned. How much did Todd earn from his summer job?

4. Mr. Maxwell started three separate bank accounts for his three children, Jerry, Hector, and Nina. He put the same amount of money in each child's account. If Tony withdrew half of this money and spent it all on a $15 CD, how much money did Mr. Maxwell deposit in total?

Solve each equation. Check your solution.

5. $9x = 72$

6. $-39.75 = 7.5m$

7. $4.8x + 7.48 = -5$

8. $\frac{8}{11}x = -\frac{4}{11}$

9. $-0.5(x - 4.2) = 6.4$

10. $-5 + 18a = -77$

11. $12 = 3.1y - 59.3$

12. $\frac{x}{5} - 9 = -4$

13. $5x + 20 = -5x$

14. $x - 19 = 2x + 12$

15. $28 - 9x = -12 - 11x$

Course 2 • Chapter 6 Equations and Inequalities

Test, Form 3B (continued)

Express an equivalent equation for each equation.

16. $-6x - 10 = -34$

16. _____

17. $-\frac{4}{5}(x + 8) = 12$

17. _____

Solve each inequality. Graph the solution on a number line.

18. $x + 3 > 13$

18. _____
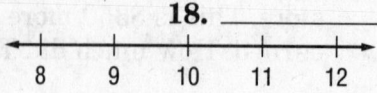

19. $n - 8 \leq 4$

19. _____
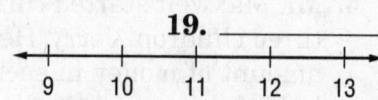

20. $9a \geq -36$

20. _____
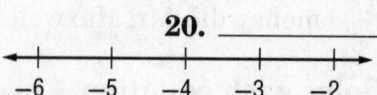

21. $-6m < -24$

21. _____
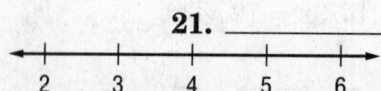

22. $\frac{w}{-2} < 8$

22. _____
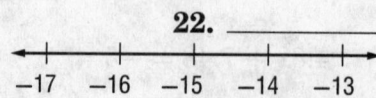

23. $\frac{b}{4} + 6 \leq 3$

23. _____
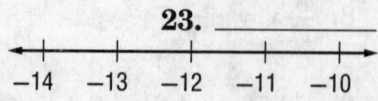

Write and solve an equation for Exercise 24.

24. The length of each side of a regular pentagon is increased by 8 inches, so the perimeter is now 65 inches. What is the original length of each side of the regular pentagon?

24. _____

25. Seth earns $7 per hour working at the library. Write and solve an inequality that can be used to find how many hours he must work in a week to earn at least $175.

25. _____

26. It costs Felisa $0.20 to send a text message from her cell phone. She has already spent $5 in text messages this month. If she has a total of $12 that she can spend this month on text messages, write and solve an inequality that will give the greatest number of text messages that she can send. Interpret the solution.

26. _____

Are You Ready?

Review

Example 1

Use a protractor to measure the angle.
Align the center of the protractor with the vertex of the angle.
Make sure one ray of the angle passes through zero on the protractor.
Read the measure on the protractor where the other ray crosses the protractor.

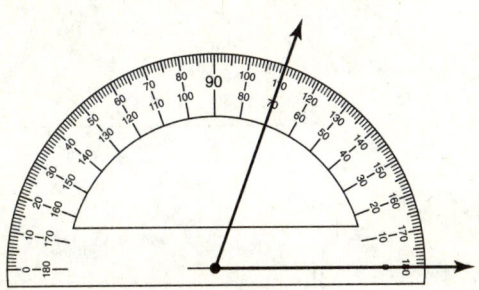

The angle measures 70°.

Example 2

Find the area of the triangle.

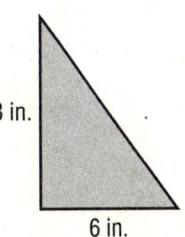

$A = \frac{1}{2} bh$ Area of a triangle.

$A = \frac{1}{2}(6 \cdot 8)$ Replace b with 6 and h with 8.

$A = 24$ Simplify.

The area of the triangle is 24 square inches.

Use a protractor to measure each angle.

1.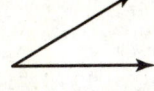

2.

1. _____

2. _____

Find the area of each triangle.

3.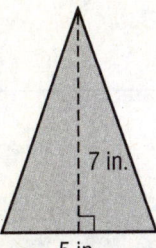

4.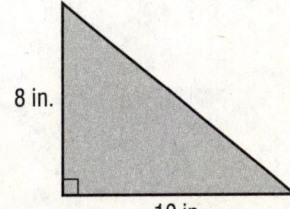

3. _____

4. _____

Course 2 • Chapter 7 Geometric Figures

145

Are You Ready?

Practice

Use a protractor to measure each angle.

1.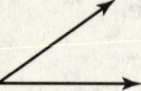

2.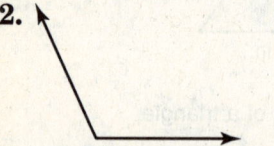

3.

4.

1. _____

2. _____

3. _____

4. _____

Find the area of each triangle.

5.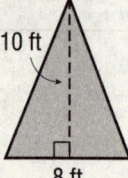
   10 ft
   8 ft

6. base: 6.5 in.
   height: 8 in.

7. base: 12 cm
   height: 7.5 cm

5. _____

6. _____

7. _____

Are You Ready?

Apply

1. **GARDENS** Zack is planting a garden that is in the shape of a triangle. The base of the triangle is 12 feet and the height is 11 feet. What is the area of the garden?

2. **CONSTRUCTION** A builder constructs a deck that contains the angle shown. Use a protractor to find the measure of the angle.

3. **SEWING** Ava cut out a triangle for her quilt. The base of the triangle is 8 inches and the height is 7.5 inches. What is the area of the triangle?

4. **ART** A piece of artwork is made from series of interlocking angles. One of the angles is shown below. Use a protractor to find the measure of the angle.

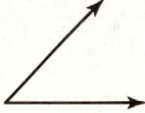

5. **PONDS** A reflecting pond is in the shape of a triangle. It has a base of 21 feet and a height of 15 feet. What is the area of the pond?

6. **ROADS** Two roads intersect to form the angle shown below. Use a protractor to find the measure of the angle.

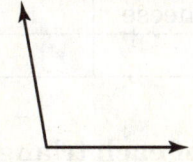

Course 2 • Chapter 7 Geometric Figures

Diagnostic Test

Use a protractor to measure each angle.

1.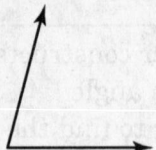
2.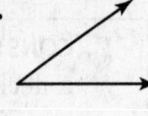

3.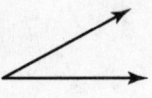
4.

5.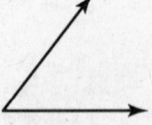
6.

7.
8.

1. _____
2. _____
3. _____
4. _____
5. _____
6. _____
7. _____
8. _____

9. **FUNDRAISING** There were 360 orders for popcorn tins, as shown in the table. How many orders were for caramel popcorn?

Popcorn Flavor	Number of Tins
butter	118
caramel	■
cheddar cheese	98
spicy	42

Find the area of each triangle.

10. base: 4.5 yd
    height: 2 yd

11. base: 3 cm
    height: 6 cm

9. _____

10. _____

11. _____

148   Course 2 • Chapter 7 Geometric Figures

Pretest

Classify each pair of angles as *complementary*, *supplementary*, or *neither*.

1.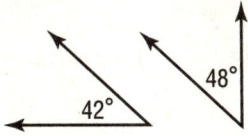

2.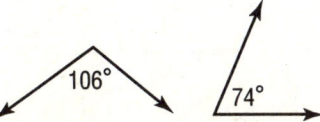

3. Classify the triangle as *acute*, *right*, or *obtuse*.

   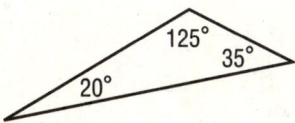

4. Classify the triangle as *scalene*, *isosceles*, or *equilateral*.

   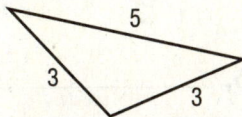

Find the value of *x* in each figure.

5.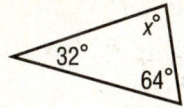

6.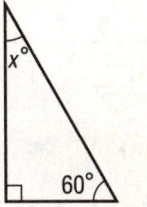

7.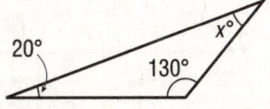

8. A model of a treehouse is made using a scale of 1 inch: 3 feet. What is the height of the actual treehouse if the height of the model is $5\frac{1}{3}$ inches?

1. _____

2. _____

3. _____

4. _____

5. _____

6. _____

7. _____

8. _____

Course 2 • Chapter 7 Geometric Figures

NAME _____ DATE _____ PERIOD _____

Chapter Quiz

Identify a pair of vertical angles.

1.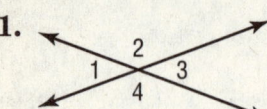

1. _____

Classify each angle as *acute, obtuse, right*, or *straight*.

2.    3.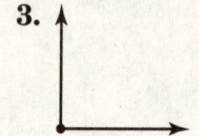

2. _____

3. _____

Classify each pair of angles as *complementary, supplementary*, or *neither*.

4. 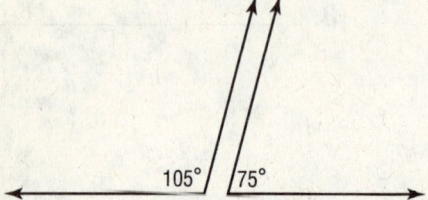   5.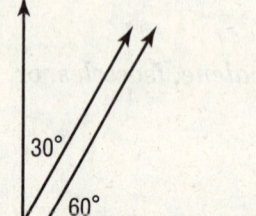

4. _____

5. _____

Classify each triangle as *acute, right*, or *obtuse*. Then find the value of x in each figure.

6.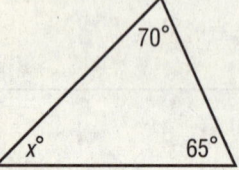

6. _____

7.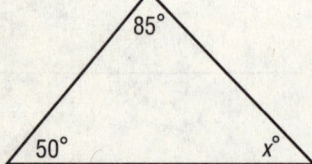

7. _____

8. **PLAYGROUND** The school playground is in the shape of a pentagon. There is a drinking fountain at each of the 5 corners of the playground. How many ways can someone walk from one drinking fountain to another drinking fountain?

8. _____

150   Course 2 • Chapter 7 Geometric Figures

NAME _____ DATE _____ PERIOD _____

Vocabulary Test

SCORE _____

acute angle	edge	scale
acute triangle	equilateral triangle	scale drawing
adjacent angles	face	scale factor
base	isosceles triangle	scale model
complementary angles	obtuse angle	scalene triangle
cone	obtuse triangle	straight angles
congruent	plane	supplementary angles
congruent segments	polyhedron	triangle
coplanar	prism	vertex
cross section	pyramid	vertical angles
cylinder	right angle	
diagonal	right triangle	

Choose from the terms above to complete each sentence.

1. The sum of the measures of _____ is 90°.

    1. _____

2. A(n) _____ measures between 90° and 180°.

    2. _____

3. A(n) _____ is a figure with three sides and three angles.

    3. _____

4. The _____ gives the ratio that compares the measurements of the drawing or model to the measurements of the real object.

    4. _____

5. A(n) _____ is a three-dimensional figure with one base that is a polygon.

    5. _____

6. Sides with the same length are _____.

    6. _____

7. A scale written as a ratio without units in simplest form is called the _____.

    7. _____

8. A(n) _____ is a flat surface that goes on forever in all directions.

    8. _____

Course 2 • Chapter 7 Geometric Figures

Standardized Test Practice

Read each question. Then fill in the correct answer on the answer document provided by your teacher or on a sheet of paper.

1. Which of the following two angles are complementary?

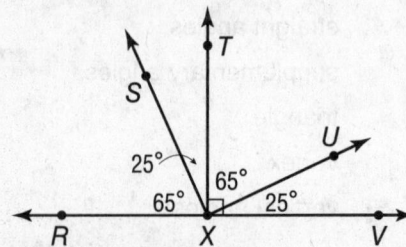

   A. ∠RXS and ∠TXU
   B. ∠SXT and ∠TXU
   C. ∠RXS and ∠SXV
   D. ∠SXR and ∠UXV

2. A recipe calls for $2\frac{1}{3}$ packages of pudding. How many batches can be made if 20 packages of pudding are available?
   F. 8 batches
   G. 9 batches
   H. 10 batches
   I. 11 batches

3. **GRIDDED RESPONSE** Greta packs tomatoes in boxes that weigh 1.4 kilograms when empty. The average tomato weighs 0.2 kilogram, and the total weight of a box filled with tomatoes is 11 kilograms. How many tomatoes are packed in each box?

4. What is the solution of the inequality below?
$$4n - 8 \leq 40$$
   A. $n \leq 8$ C. $n \geq 8$
   B. $n \leq 12$ D. $n \geq 12$

5. In the figure below, line x is parallel to line y.

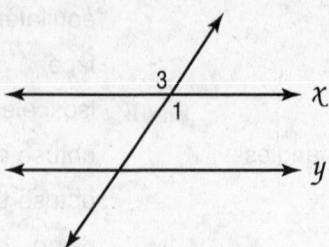

   What type of angles are ∠1 and ∠3?
   F. vertical angles
   G. adjacent angles
   H. right angles
   I. regular angles

6. Thom has a scale model of his car. The scale is 1 : 12. If the actual car has 16-inch wheels, what size are the wheels on the scale model?

   A. $1.\overline{3}$ in.
   B. 2 in.
   C. 32 in.
   D. 40 in.

7. **SHORT RESPONSE** In triangle ABC, $m\angle A = 55°$ and $m\angle B = 35°$. Classify the triangle by its angles.

8. A stained glass window is in the shape of an equilateral triangle. What is the measure of one interior angle of the triangle?
   F. 30°
   G. 60°
   H. 90°
   I. 180°

9. **SHORT RESPONSE** In △FGH, $m\angle G = 30°$ and $m\angle H = 100°$. What is the measure of $\angle F$?

10. What is the measure of ∠1 in the figure?

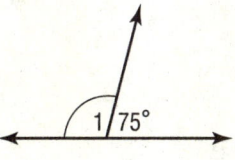

   A. 15° C. 100°
   B. 25° D. 105°

11. The bridge structure is supported by the triangular braces as shown.

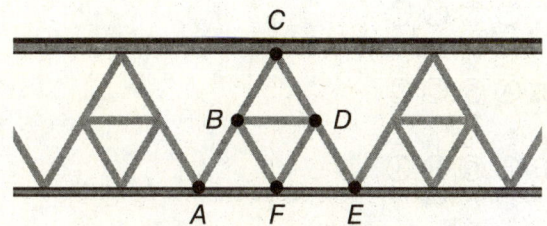

   Triangles ACE and ABF are similar triangles. The scale factor is 0.5. If CE = 10 feet, what is the length of BF?
   F. 2.5 ft H. 6 ft
   G. 5 ft   I. 12 ft

12. Jesse purchased a new digital camera for $499 and a printer for $299 including tax. He plans to pay the total amount in 6 equal monthly payments. What is a reasonable estimate of the amount he will pay each month?
   A. $66.50   C. $155.00
   B. $133.00 D. $165.00

13. An architect created the scale drawing below showing a wall of a child's playhouse.

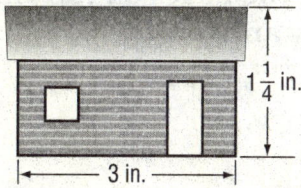

   Which of these was the scale used for the drawing if the actual height of the wall is $7\frac{1}{2}$ feet?
   F. 1 in. = 1 ft
   G. $\frac{1}{2}$ in. = 1 ft
   H. 2 in. = 12 ft
   I. $\frac{1}{4}$ in. = 1 ft

14. **EXTENDED RESPONSE** Use triangle XYZ to answer the following questions.

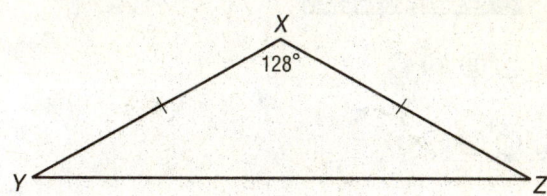

   Part A Classify angle X.
   Part B Classify angle Y.
   Part C Classify the triangle by its sides and by its angles.
   Part D If ∠Y is congruent to ∠Z, find the measure of ∠Z. Explain.

NAME _____ DATE _____ PERIOD _____

Student Recording Sheet

SCORE _____

Use this recording sheet with the Standardized Test Practice pages.

Fill in the correct answer. For gridded-response questions, write your answers in the boxes on the answer grid and fill in the bubbles to match your answers.

1. Ⓐ Ⓑ Ⓒ Ⓓ

2. Ⓕ Ⓖ Ⓗ Ⓘ

3. [grid-in response]

4. Ⓐ Ⓑ Ⓒ Ⓓ

5. Ⓕ Ⓖ Ⓗ Ⓘ

6. Ⓐ Ⓑ Ⓒ Ⓓ

7. _____

8. Ⓕ Ⓖ Ⓗ Ⓘ

9. _____

10. Ⓐ Ⓑ Ⓒ Ⓓ

11. Ⓕ Ⓖ Ⓗ Ⓘ

12. Ⓐ Ⓑ Ⓒ Ⓓ

13. Ⓕ Ⓖ Ⓗ Ⓘ

Extended Response
Record your answers for Exercise 14 on the back of this paper.

Course 2 • Chapter 7 Geometric Figures

NAME _____ DATE _____ PERIOD _____

Extended-Response Test

SCORE _____

Demonstrate your knowledge by giving a clear, concise solution to each problem. Be sure to include all relevant drawings and justify your answers. You may show your solutions in more than one way or investigate beyond the requirements of the problem. If necessary, record your answer on another piece of paper.

1. **a.** Explain what it means for two angles to be congruent.
   b. In the figure, which angles are congruent? Find the measure of all the angles if $m\angle 2 = 76°$.

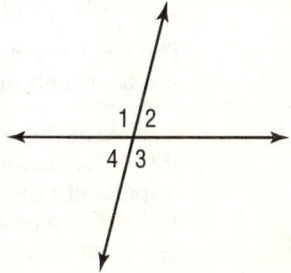

2. **a.** Explain how to classify triangles by the measure of their angles. Draw an example of each classification.
   b. Draw a right angle, obtuse angle, straight angle, and acute angle. Explain how the figures are different.
   c. Draw an example of supplementary angles. Explain how supplementary angles are different from complementary angles.

3. **a.** Explain how to classify the figure at the right.
   b. Name the base(s). Explain how you determined your response.

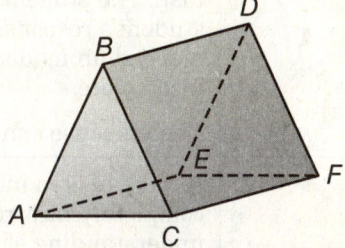

Course 2 • Chapter 7 Geometric Figures

155

NAME _____ DATE _____ PERIOD _____

Extended-Response Rubric

SCORE _____

Score	Description
4	A score of four is a response in which the student demonstrates a thorough understanding of the mathematics concepts and/or procedures embodied in the task. The student has responded correctly to the task, used mathematically sound procedures, and provided clear and complete explanations and interpretations. The response may contain minor flaws that do not detract from the demonstration of a thorough understanding.
3	A score of three is a response in which the student demonstrates an understanding of the mathematics concepts and/or procedures embodied in the task. The student's response to the task is essentially correct with the mathematical procedures used and the explanations and interpretations provided demonstrating an essential but less than thorough understanding. The response may contain minor flaws that reflect inattentive execution of mathematical procedures or indications of some misunderstanding of the underlying mathematics concepts and/or procedures.
2	A score of two indicates that the student has demonstrated only a partial understanding of the mathematics concepts and/or procedures embodied in the task. Although the student may have used the correct approach to obtaining a solution or may have provided a correct solution, the student's work lacks an essential understanding of the underlying mathematical concepts. The response contains errors related to misunderstanding important aspects of the task, misuse of mathematical procedures, or faulty interpretations of results.
1	A score of one indicates that the student has demonstrated a very limited understanding of the mathematics concepts and/or procedures embodied in the task. The student's response is incomplete and exhibits many flaws. Although the student's response has addressed some of the conditions of the task, the student reached an inadequate conclusion and/or provided reasoning that was faulty or incomplete. The response exhibits many flaws or may be incomplete.
0	A score of zero indicates that the student has provided no response at all, or a completely incorrect or uninterpretable response, or demonstrated insufficient understanding of the mathematics concepts and/or procedures embodied in the task. For example, a student may provide some work that is mathematically correct, but the work does not demonstrate even a rudimentary understanding of the primary focus of the task.

NAME _____ DATE _____ PERIOD _____

SCORE _____

Test, Form 1A

Write the letter for the correct answer in the blank at the right of each question.

1. What is the value of x in the figure?

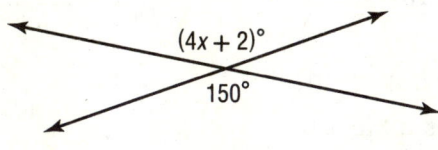

   A. 37   B. 60   C. 160   D. 237

   1. _____

2. What is the classification of the pair of angles shown?

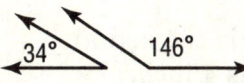

   F. complementary   H. supplementary
   G. vertical   I. no relationship

   2. _____

3. Which name applies to the angle pair at the right?
   A. supplementary
   B. straight
   C. complementary
   D. vertical

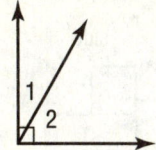

   3. _____

4. The angles shown are supplementary. What is the value of x?
   F. 180
   G. 95
   H. 27
   I. 17

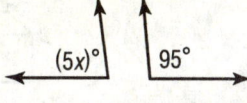

   4. _____

5. Find the value of x in the figure shown.
   A. 20
   B. 22
   C. 45
   D. 80

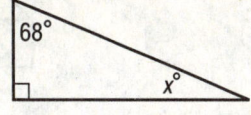

   5. _____

6. In the figure at the right, what is m∠1 if m∠3 = 70°?
   F. 180°   H. 90°
   G. 110°   I. 70°

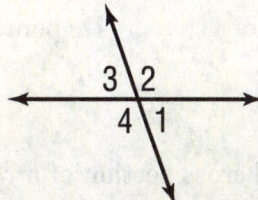

   6. _____

Course 2 • Chapter 7 Geometric Figures

Test, Form 1A   *(continued)*

7. What is the classification of the triangle by its angles and by its sides?
   A. acute, equilateral   C. obtuse, isosceles
   B. right, equilateral   D. obtuse, equilateral

   7. _____

8. On a scale drawing, the scale is $\frac{1}{2}$ inch = 1 foot. What are the dimensions on the scale drawing for a room that is 22 feet by 17 feet?
   F. $\frac{17}{12}$ in. by $9\frac{1}{2}$ in.   H. $5\frac{1}{2}$ in. by $4\frac{1}{4}$ in.
   G. $1\frac{5}{17}$ in. by $1\frac{1}{3}$ in.   I. 11 in. by $8\frac{1}{2}$ in.

   8. _____

9. On a map, the scale is 1 inch = 125 miles. What is the actual distance between two cities if the map distance is 4 inches?
   A. 525 mi   C. 500 mi
   B. 505 mi   D. $281\frac{1}{4}$ mi

   9. _____

10. Which solid has the top, the side, and the front views given?

    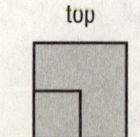 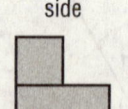 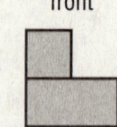
    top   side   front

    F. 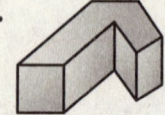   H.

    G.    I.

    10. _____

Use the figure shown on the right.

11. What is the classification of the figure?
    A. rectangular pyramid   C. rectangular prism
    B. pentagonal prism   D. pentagonal pyramid

    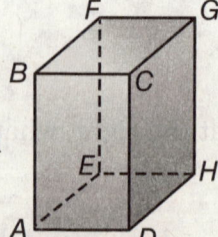

    11. _____

12. What would be the cross section of a rectangular prism that is sliced horizontally?
    F. rectangle   G. triangle   H. cirlce   I. trapezoid

    12. _____

158   Course 2 • Chapter 7 Geometric Figures

NAME _____ DATE _____ PERIOD _____

SCORE _____

Test, Form 1B

Write the letter for the correct answer in the blank at the right of each question.

1. What is the value of x in the figure?

   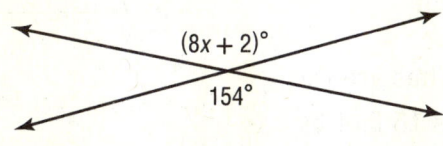

   A. 180   B. 91   C. 90   D. 19

   1. _____

2. What is the classification of the pair of angles shown?

   F. complementary   H. vertical
   G. supplementary   I. no relationship

   2. _____

3. Which name does *not* apply to the angle pair?

   A. vertical   C. supplementary
   B. adjacent   D. right angles

   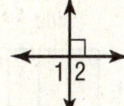

   3. _____

4. The angles shown are supplementary. What is the value of x?

   F. 180   H. 28
   G. 94   I. 17

   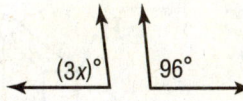

   4. _____

5. Find the value of x in the figure shown.

   A. 24
   B. 25
   C. 45
   D. 80

   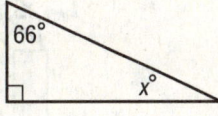

   5. _____

6. In the figure at the right, what is $m\angle 2$ if $m\angle 4 = 120°$?

   F. 60°   H. 120°
   G. 90°   I. 150°

   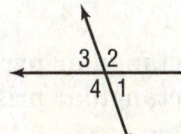

   6. _____

Course 2 • Chapter 7 Geometric Figures

Test, Form 1B (continued)

7. What is the classification of the triangle by its angles and by its sides?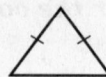
   A. right, isosceles   C. obtuse, isosceles
   B. acute, isosceles   D. acute, equilateral

   7. _____

8. On a scale drawing, the scale is $\frac{1}{4}$ inch = 1 foot. What are the dimensions on the scale drawing for a room that is 15 feet by 16 feet?
   F. $3\frac{3}{4}$ in. by 4 in.   H. $5\frac{1}{4}$ in. by $3\frac{1}{4}$ in.
   G. $4\frac{3}{4}$ in. by $2\frac{1}{3}$ in.   I. 8 in. by $8\frac{1}{2}$ in.

   8. _____

9. On a map, the scale is 1 inch = 50 miles. What is the actual distance between two cities if the map distance is 5 inches?
   A. 10 miles   C. 250 miles
   B. 25 miles   D. 300 miles

   9. _____

10. Which solid has the top, the side, and the front views given?

    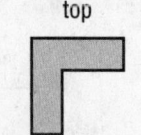 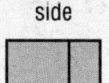 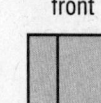

    F. 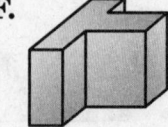   H.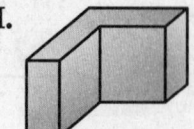

    G. 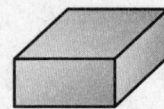   I.

    10. _____

Use the figure shown on the right.

11. What is the classification of the figure?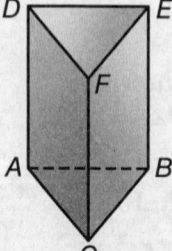
    A. triangular prism   C. rectangular pyramid
    B. triangular pyramid   D. rectangular prism

    11. _____

12. What would be the cross section of a rectangular prism that is sliced horizontally?
    F. trapezoid   G. triangle   H. rectangle   I. circle

    11. _____

NAME _____ DATE _____ PERIOD ___

Test, Form 2A

SCORE ___

Write the letter for the correct answer in the blank at the right of each question.

1. What is the classification of the angle shown?

   A. acute   C. straight
   B. right   D. obtuse

   1. _____

2. What is the value of x?

   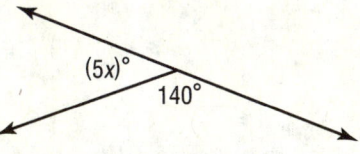

   F. 180   H. 40
   G. 90   I. 8

   2. _____

3. What is the value of x in the figure at the right?
   A. 90   C. 120
   B. 60   D. 30

   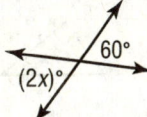

   3. _____

4. Angle 1 and angle 2 are supplementary. If $m\angle 1 = 27°$, what is $m\angle 2$?
   F. 27°   H. 153°
   G. 63°   I. 163°

   4. _____

5. A building is 120 meters tall. A scale model of the building uses a scale of 1 centimeter = 6 meters. How tall is the model?
   A. 20 cm   C. 20 m
   B. 60 cm   D. 60 m

   5. _____

6. Which pair of angles is congruent?
   F. ∠1 and ∠2
   G. ∠1 and ∠4
   H. ∠4 and ∠3
   I. ∠4 and ∠2

   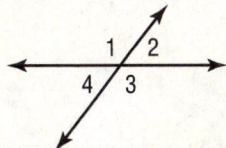

   6. _____

Course 2 • Chapter 7 Geometric Figures

161

Test, Form 2A (continued)

7. On a scale drawing, $\frac{1}{4}$ inch = 1 foot. What are the dimensions on the scale for a room that is 27 feet by 20 feet?

7. _____

8. What is the value of x in the triangle?

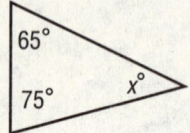

8. _____

9. Draw a top, a side, and a front view of the solid.

9. _____

10. What is the classification of the triangle by its angles and by its sides?

10. _____

Use the figure shown at the right.

11. What is the name of the figure?

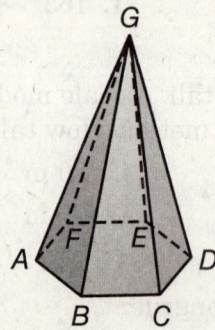

11. _____

12. What would be the cross section from a vertical slice of the figure in Exercise 11?

12. _____

NAME _____ DATE _____ PERIOD _____

Test, Form 2B

SCORE _____

Write the letter for the correct answer in the blank at the right of each question.

1. What is the classification of the angle shown?

   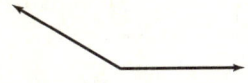

   A. acute **B.** right **C.** straight **D.** obtuse

   1. _____

2. What is the value of x?

   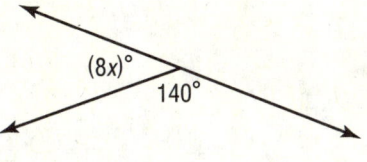

   F. 180 **H.** 45
   G. 90   **I.** 5

   2. _____

3. What is the value of x in the figure at the right?

   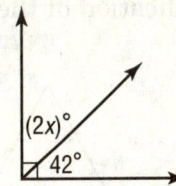

   A. 9 **B.** 24 **C.** 139 **D.** 141

   3. _____

4. Angle 1 and angle 2 are supplementary. If $m\angle 1 = 63°$, what is $m\angle 2$?
   F. 27° **G.** 63° **H.** 117° **I.** 153°

   4. _____

5. On a map, the scale is 1 inch = 125 miles. What is the actual distance between the two cities if the map distance is $2\frac{3}{4}$ inches?

   A. 257 mi **B.** $281\frac{1}{4}$ mi **C.** 325 mi **D.** $343\frac{3}{4}$ mi

   5. _____

6. Which pair of angles is congruent?
   F. $\angle 8$ and $\angle 7$
   G. $\angle 8$ and $\angle 5$
   H. $\angle 5$ and $\angle 6$
   I. $\angle 7$ and $\angle 5$

   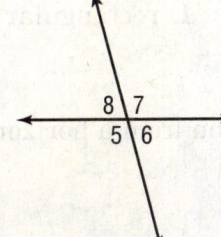

   6. _____

Course 2 • Chapter 7 Geometric Figures

Test, Form 2B (continued)

7. On a scale drawing, the scale is $\frac{1}{4}$ inch = 1 foot. What are the dimensions on the scale drawing for a room that is 18 feet by 16 feet?

   7. _____

8. What is the value of x in the triangle?

   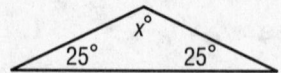

   8. _____

9. Draw a top, a side, and a front view of the solid.

   9. _____

10. What is the classification of the triangle by its angles and by its sides?

    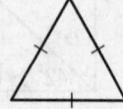

    10. _____

Use the figure below to answer Exercises 11 and 12.

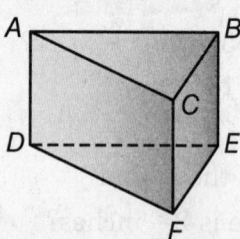

11. What is the name of the figure?
    - F. triangular pyramid
    - G. hexagonal prism
    - H. triangular prism
    - I. rectangular prism

    11. _____

12. What would be the cross section from a horizontal slice of the figure shown in Exercise 11?

    12. _____

NAME _____ DATE _____ PERIOD _____

Test, Form 3A

SCORE _____

Find the value of x in each figure.

1.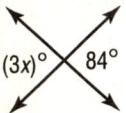

2.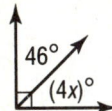

3.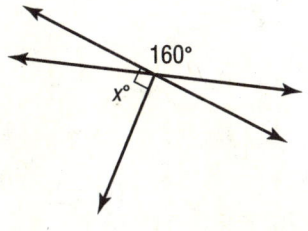

4.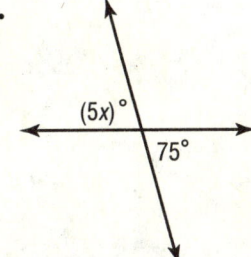

1. _____

2. _____

3. _____

4. _____

5. Suppose you are making a scale drawing. Find the length of each object on the scale drawing with the given scale. Then find the scale factor.
   a. a room 14 feet long; 1 inch = 6 feet
   b. a tower 40 meters tall; 0.8 centimeter = 1 meter.

5a. _____

5b. _____

Find the missing measure in each triangle. Then classify the triangle as *acute*, *right*, or *obtuse*.

6.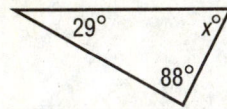

7.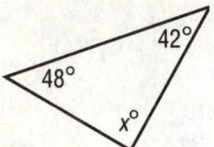

6. _____

7. _____

8. A model of a building is made using a scale of 1 inch = 20 feet. What is the height of the actual building if the height of the model is 10.5 inches?

8. _____

Course 2 • Chapter 7 Geometric Figures

NAME _____ DATE _____ PERIOD _____

Test, Form 3A (continued)

SCORE _____

Draw a top, a side, and a front view of each solid.

9.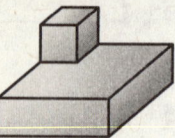

9. _____

10.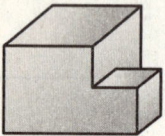

10. _____

11. Identify the figure. Then name the bases, faces, edges, and vertices.

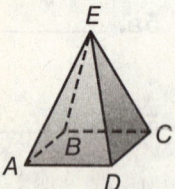

11. _____

12. Describe the shape that would result from a vertical slice of the figure below.

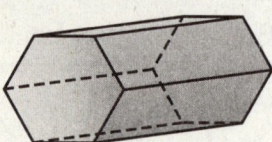

12. _____

13. Explain how a cross section of the figure below can result in a trapezoid.

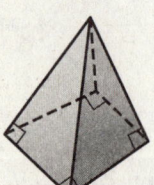

13. _____

166 Course 2 • Chapter 7 Geometric Figures

NAME _____ DATE _____ PERIOD _____

Test, Form 3B

SCORE _____

Find the value of x in each figure.

1.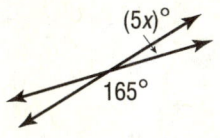

2.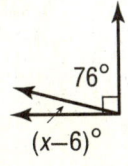

3. (figure with 20° and $(x+15)°$ forming a right angle)

4. (figure showing $(3x)°$ at a right angle with perpendicular lines)

1. _____

2. _____

3. _____

4. _____

5. Suppose you are making a scale drawing. Find the length of each object on the scale drawing with the given scale. Then find the scale factor.
   a. a parking lot 480 meters wide; 1 centimeter = 16.5 meters
   b. a desk 6 feet long; 1.5 inches = 0.5 feet

5a. _____

5b. _____

Find the missing angle measure in each triangle. Then classify the triangle as *acute*, *right*, or *obtuse*.

6.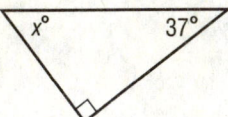

7.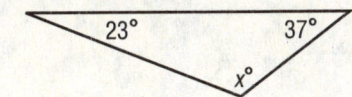

6. _____

7. _____

8. A model of a building is made using a scale of 1 inch = 25 feet. What is the height of the actual building if the height of the model is 12.5 inches?

8. _____

Course 2 • Chapter 7 Geometric Figures

167

Test, Form 3B (continued)

Draw a top, a side, and a front view of each solid.

9.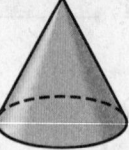

9. _____

10.

10. _____

11. Identify the figure. Then name the bases, faces, edges, and vertices.

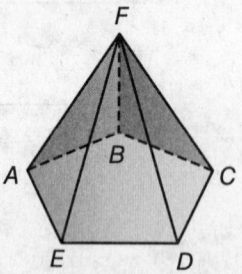

11. _____

12. Describe the shape that would result from a horizontal slice of the figure below.

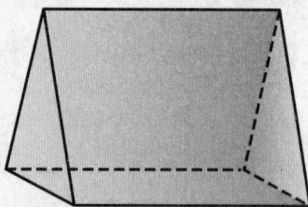

12. _____

13. Explain how the horizontal, vertical, and angled cross sections of the figure shown are related.

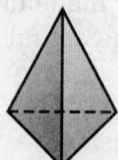

13. _____

168 Course 2 • Chapter 7 Geometric Figures

NAME _____ DATE _____ PERIOD _____

Are You Ready?

Review

Example 1
Find the area of the rectangle.

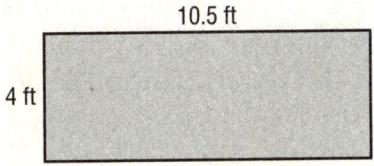

$A = lw$ Area of a rectangle.
$A = (10.5)(4)$ Replace *l* with 10.5 and *w* with 4.
$A = 42$ Simplify.

The area of the rectangle is 42 square feet.

Example 2
Find the area of the triangle.

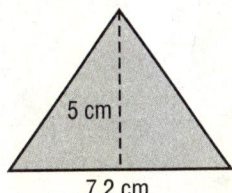

$A = \frac{1}{2}bh$ Area of a triangle.
$A = \frac{1}{2}(7.2)(5)$ Replace *b* with 7.2 and *h* with 5.
$A = 36$ Simplify.

Exercises

1. Find the area of the rectangle.

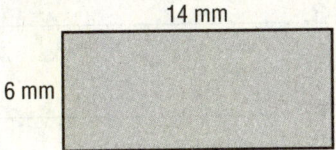

2. Find the area of the triangle.

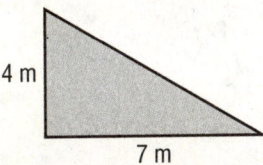

3. Find the area of the triangle.

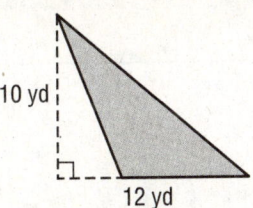

4. Find the area of the rectangle.

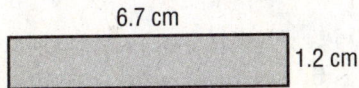

5. **PAPER** A typical page of notebook paper is 11 inches long by 8.5 inches wide. What is the area of a typical page of notebook paper?

1. _____
2. _____
3. _____
4. _____
5. _____

Course 2 • Chapter 8 Measure Figures   169

Are You Ready?

Practice

Find the area of each rectangle.

1.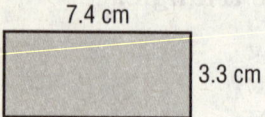

2.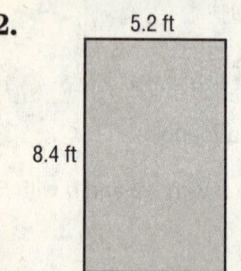

3.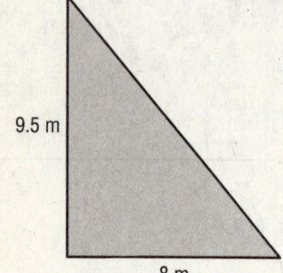

4. Find the area of a rectangle that has a length of 17 inches and a width of 7 inches.

Find the area of each triangle

5.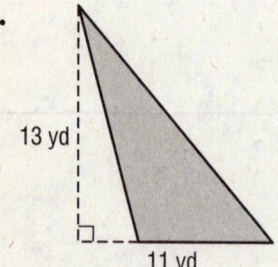

6.

7. Find the area of a triangle that has a base of 23 inches and a height of 25 inches.

8. **SANDBOX** A sandbox at a playground is 3 yards long by 2.3 yards wide. What is the area of the sandbox?

1. _____
2. _____
3. _____
4. _____
5. _____
6. _____
7. _____
8. _____

NAME _____ DATE _____ PERIOD _____

Are You Ready?

Apply

1. **FARMING** A farmer has a rectangular field that is 100 yards long by 20 yards wide. What is the total area of the field?

2. **WALLPAPER** Sydney is going to wallpaper the rectangular wall in her bedroom. If her wall is 16 feet long by 8 feet tall, how much wallpaper will she need?

3. **GARDENING** Elena wants to put a new layer of soil in her triangular garden. She needs to know the area of her garden so that she knows how much soil to buy. If the garden has a base of 14.3 meters and a height of 6 meters, how much area does she need to cover?

4. **TILES** Mr. McCabe is buying tiles for his rectangular kitchen floor. The floor is 15 feet long by 25 feet wide. If each tile is 1 square foot, how many tiles does he need?

5. **ARTS** Ariel is making a sign in the shape of a triangle. The triangle has a base of 30.5 inches and a height of 36 inches. What is the area of the triangle?

6. **WILDFIRES** A spokesperson from the forest fire service said that a wildfire caused a rectangular patch of forest that was 26 miles long by 24 miles wide to be burned. How many square miles of forest were burned by the wildfire?

Course 2 • Chapter 8 Measure Figures

Diagnostic Test

Find the area of each rectangle.

1.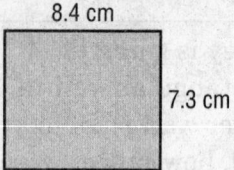

2.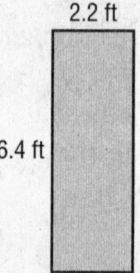

3.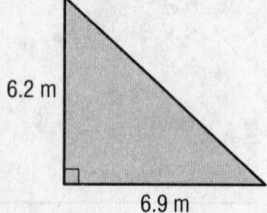

1. _____

2. _____

3. _____

Find the area of each triangle.

4.

5.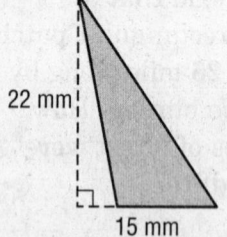

4. _____

5. _____

6. Find the area of a triangle that has a base of 50 inches and a height of 35 inches.

6. _____

7. Find the area of a rectangle that has a length of 16 inches and a width of 5.5 inches.

7. _____

8. **POOL COVER** A large pool cover is 80 feet long by 40 feet wide. What is the area of the pool cover?

8. _____

172   Course 2 • Chapter 8 Measure Figures

Pretest

Find the volume of each figure.

1.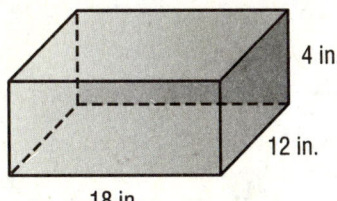

2.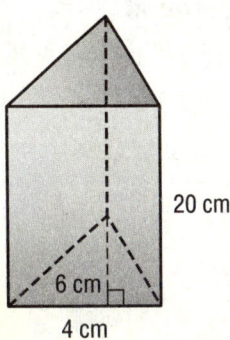

3.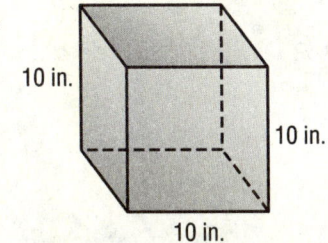

4.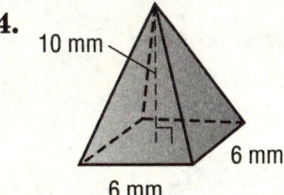

1. _____

2. _____

3. _____

4. _____

Find the surface area of each figure.

5.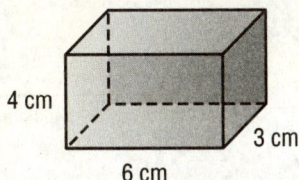

6.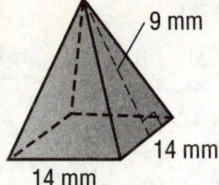

5. _____

6. _____

Course 2 • Chapter 8 Measure Figures

Chapter Quiz

Find the circumference of each circle. Round to the nearest tenth. Use 3.14 or $\frac{22}{7}$ for π.

1.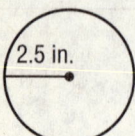

2.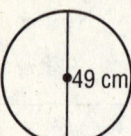

1. _____

2. _____

Find the area of each figure. Round to the nearest tenth. Use 3.14 or $\frac{22}{7}$ for π.

3.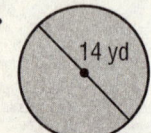

4.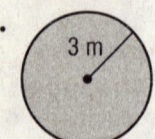

5.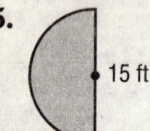

3. _____

4. _____

5. _____

Find the volume of each figure. Round to the nearest tenth if necessary.

6.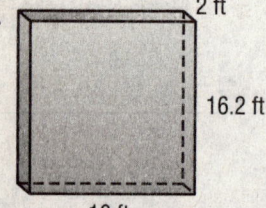

7.

6. _____

7. _____

174 Course 2 • Chapter 8 Measure Figures

NAME _____ DATE _____ PERIOD _____

Vocabulary Test

center	lateral face	regular pyramid
circle	lateral surface area	semicircle
circumference	pi	slant height
composite figure	pyramid	surface area
diameter	radius	volume

Write the letter of the term that best matches each statement or phrase. Some terms may be used more than once.

1. the distance around a circle   **a.** radius   1. _____

2. the distance across a circle through its center   **b.** volume   2. _____

3. height of each lateral face of a pyramid   **c.** pi   3. _____

4. the ratio of the circumference of a circle to its diameter   **d.** pyramid   4. _____

5. the measure of space occupied by a three-dimensional figure   **e.** composite figure   5. _____

6. the expression $2\ell w + 2wh + 2\ell h$ is used to find this measure of a rectangular prism   **f.** diameter   6. _____

7. a figure made up of two or more three-dimensional figures   **g.** surface area   7. _____

8. polyhedron with one base that is a polygon and three or more triangular faces that meet at a common vertex.   **h.** slant height   8. _____

9. the distance from the center to any point on the circle   **i.** circumference   9. _____

10. sum of the areas of all the faces of a three-dimensional figure   10. _____

11. Define lateral surface area in your own words.   11. _____

Course 2 • Chapter 8 Measure Figures

Standardized Test Practice

Read each question. Then fill in the correct answer on the answer sheet provided by your teacher or on a sheet of paper.

1. A metal toolbox has a length of 11 inches, a width of 5 inches, and a height of 6 inches. What is the volume of the toolbox?
   A. 22 in³
   B. 121 in³
   C. 210 in³
   D. 330 in³

2. Evelyn has 3 apples to serve to her friends. If Evelyn serves each friend $\frac{1}{3}$ of a whole apple, how many friends can she serve?
   F. 1
   G. 3
   H. 9
   I. 12

3. **GRIDDED RESPONSE** Daniel is designing and building a small storage shed. He wants the dimensions of the shed to be one half the dimensions of the shed shown below.

   Storage Shed

   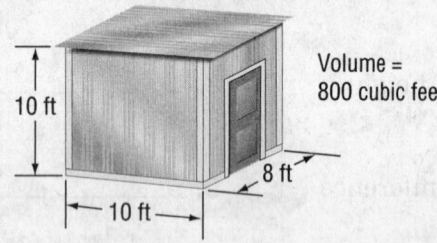

   If the dimensions of the shed above are each divided in half, the volume of Daniel's new storage shed will be what fraction of the volume of the original storage shed?

4. **GRIDDED RESPONSE** Timea ran 3 miles in 19 minutes. At this rate, how many minutes would it take her to run 5 miles?

5. Wilma made a decorative piece shaped like a square pyramid with the dimensions shown.
   She wants to double the volume of the piece. Which of the following square pyramid pieces will have a volume that is twice the volume of Wilma's decorative piece?

   A.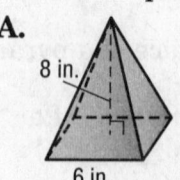
   B.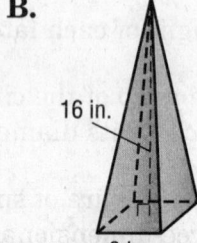

   C.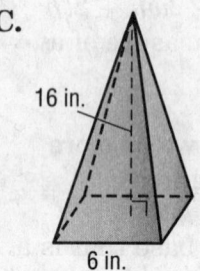
   D.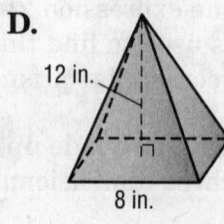

6. A wallet-sized print is about 5 centimeters wide and 8 centimeters long. Grace wants to use the wallet-sized print to make a print that is similar to the wallet-sized print. If the new print will be 20 centimeters long, how wide will the new print be to the nearest centimeter?
   F. 11 centimeters
   G. 12 centimeters
   H. 13 centimeters
   I. 15 centimeters

NAME _____ DATE _____ PERIOD _____

7. **GRIDDED RESPONSE** Solve the equation $b - 5 = -8$. What is the value of b?

8. What is the area of the circle shown below?

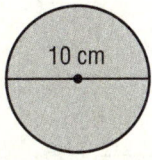

   A. 314 cm²
   B. 78.5 cm²
   C. 15.7 cm³
   D. 3.14 cm²

9. **SHORT RESPONSE** Compare the surface area of the figures. Justify your answer.

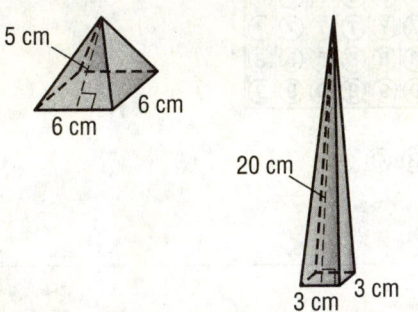

10. The circumference of a circle is 37.68 meters. What is the radius of the circle?

    F. 12 m
    G. 10 m
    H. 6 m
    I. 3.14 m

11. **SHORT RESPONSE** Sophie drew a circle on a map with a radius of 2 inches. She plans to visit the cities within the circle. What is the area of the map that she wants to visit?

12. Andrea made a tiered cake for a wedding. She wants to cover the outside of each layer marked A, B, and C with white icing.

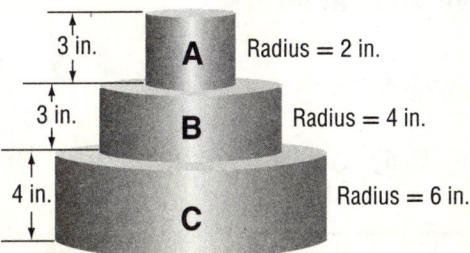

What is the circumference of cake A?

A. 3.14 in.
B. 12.56 in.
C. 25.12 in.
D. 37.68 in.

13. **EXTENDED RESPONSE** A ceramic dish company makes small square dishes with lids. The dishes are shipped in rectangular boxes that are 20 centimeters by 20 centimeters by 16 centimeters. The extra space in the box is filled with packing material to protect the dish.

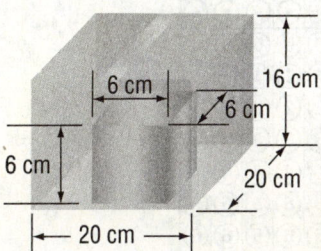

Part A How much space does the dish take up?

Part B How much packing material is needed?

Part C How much material is needed to make the box?

Course 2 • Chapter 8 Measure Figures   **177**

NAME _____ DATE _____ PERIOD _____

Student Recording Sheet

SCORE _____

Use this recording sheet with the Standardized Test Practice pages.

Fill in the correct answer. For gridded-response questions, write your answers in the boxes on the answer grid and fill in the bubbles to match your answers.

1. Ⓐ Ⓑ Ⓒ Ⓓ

2. Ⓕ Ⓖ Ⓗ Ⓘ

3. [gridded response]

4. [gridded response]

5. Ⓐ Ⓑ Ⓒ Ⓓ

6. Ⓕ Ⓖ Ⓗ Ⓘ

7. [gridded response]

8. Ⓐ Ⓑ Ⓒ Ⓓ

9. _____

10. Ⓕ Ⓖ Ⓗ Ⓘ

11. _____

12. Ⓐ Ⓑ Ⓒ Ⓓ

Extended Response
Record your answers for Exercise 13 on the back of this paper.

178   Course 2 • Chapter 8 Measure Figures

Extended-Response Test

Demonstrate your knowledge by giving a clear, concise solution to each problem. Be sure to include all relevant drawings and justify your answers. You may show your solutions in more than one way or investigate beyond the requirements of the problems. If necessary, record your answers on another piece of paper.

1. **a.** Draw a diagram of a circle and its parts. Draw and label the circumference, diameter, and radius.

   b. Suppose the radius of the circle is 7 meters. Find the diameter and the circumference of the circle using the number for the radius. Estimate the circumference of the circle. Use a formula and show your work.

2. Use the boxes shown below.

   A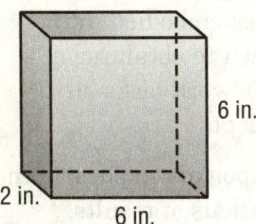

   B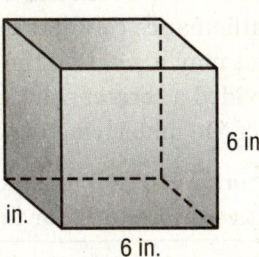

   C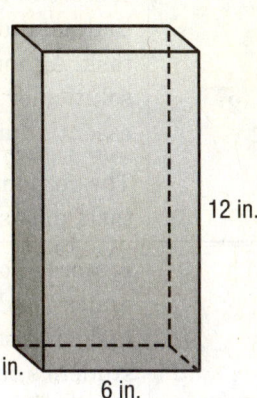

   a. Compare the volumes and surface areas of the three boxes.

   b. Explain why you think manufacturers frequently package their products in tall boxes, bottles, or cans.

Course 2 • Chapter 8 Measure Figures 179

NAME _____ DATE _____ PERIOD _____

Extended-Response Rubric

SCORE _____

Score	Description
4	A score of four is a response in which the student demonstrates a thorough understanding of the mathematics concepts and/or procedures embodied in the task. The student has responded correctly to the task, used mathematically sound procedures, and provided clear and complete explanations and interpretations. The response may contain minor flaws that do not detract from the demonstration of a thorough understanding.
3	A score of three is a response in which the student demonstrates an understanding of the mathematics concepts and/or procedures embodied in the task. The student's response to the task is essentially correct with the mathematical procedures used and the explanations and interpretations provided demonstrating an essential but less than thorough understanding. The response may contain minor flaws that reflect inattentive execution of mathematical procedures or indications of some misunderstanding of the underlying mathematics concepts and/or procedures.
2	A score of two indicates that the student has demonstrated only a partial understanding of the mathematics concepts and/or procedures embodied in the task. Although the student may have used the correct approach to obtaining a solution or may have provided a correct solution, the student's work lacks an essential understanding of the underlying mathematical concepts. The response contains errors related to misunderstanding important aspects of the task, misuse of mathematical procedures, or faulty interpretations of results.
1	A score of one indicates that the student has demonstrated a very limited understanding of the mathematics concepts and/or procedures embodied in the task. The student's response is incomplete and exhibits many flaws. Although the student's response has addressed some of the conditions of the task, the student reached an inadequate conclusion and/or provided reasoning that was faulty or incomplete. The response exhibits many flaws or may be incomplete.
0	A score of zero indicates that the student has provided no response at all, or a completely incorrect or uninterpretable response, or demonstrated insufficient understanding of the mathematics concepts and/or procedures embodied in the task. For example, a student may provide some work that is mathematically correct, but the work does not demonstrate even a rudimentary understanding of the primary focus of the task.

NAME _____ DATE _____ PERIOD _____

Test, Form 1A

SCORE _____

Write the letter for the correct answer in the blank at the right of each question.

What is the circumference of each circle? Use 3.14 for π. Round to the nearest tenth if necessary.

1. A. 7.9 mm   C. 31.4 mm
   B. 15.7 mm D. 78.5 mm

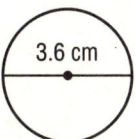

1. _____

2. F. 22.6 cm   H. 10.2 cm
   G. 11.3 cm   I. 5.7 cm

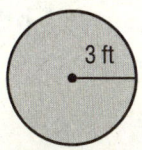

2. _____

What is the area of each circle? Use 3.14 for π. Round to the nearest tenth if necessary.

3. A. 9.4 ft²   C. 28.3 ft²
   B. 18.8 ft² D. 113 ft²

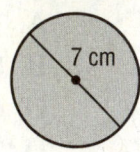

3. _____

4. F. 38.5 cm² H. 153.9 cm²
   G. 44 cm²   I. 615.4 cm²

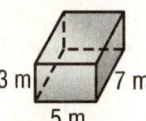

4. _____

5. What is the volume of the rectangular prism?
   A. 15 m³   C. 142 m³
   B. 105 m³   D. 210 m³

5. _____

6. A cube has 4-inch edges. What is its volume?
   F. 12 in³   G. 16 in³   H. 64 in³   I. 96 in³

6. _____

For Exercises 7 and 8, what is the volume of each figure? Round to the nearest tenth if necessary.

7.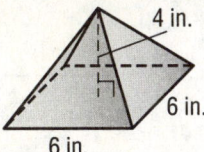
   A. 32 in³
   B. 48 in³
   C. 72 in³
   D. 144 in³

7. _____

8.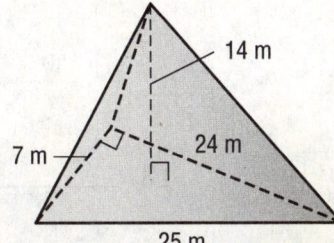
   F. 2,352 m³
   G. 1,176 m³
   H. 392 m³
   I. 261.3 m³

8. _____

Course 2 · Chapter 8 Measure Figures

NAME _____ DATE _____ PERIOD _____

Test, Form 1A (continued)   SCORE _____

For Exercises 9–11, what is the surface area of each figure? Round to the nearest tenth if necessary.

9.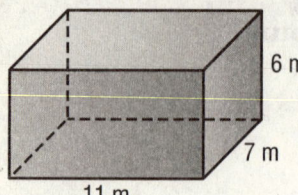

A. 185 m²
B. 231 m²
C. 370 m²
D. 462 m²

9. _____

10.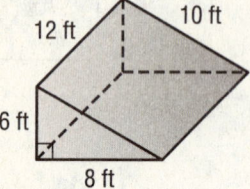

F. 134 ft²
G. 144 ft²
H. 288 ft²
I. 336 ft²

10. _____

11.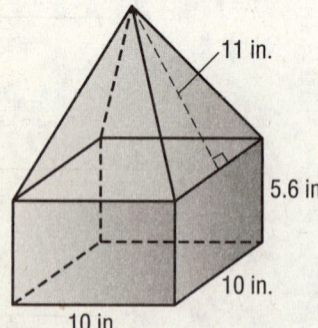

A. 444 in²
B. 544 in²
C. 644 in²
D. 780 in²

11. _____

12. Find the area of the composite figure.

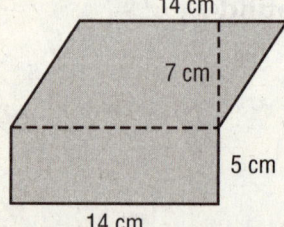

F. 168 cm²
G. 150 cm²
H. 144 cm²
I. 130 cm²

12. _____

13. Find the total surface area of the pyramid.

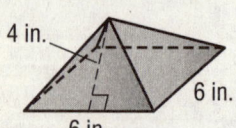

A. 84 in²
B. 90 in²
C. 100 in²
D. 130 in²

13. _____

182   Course 2 • Chapter 8 Measure Figures

NAME _____ DATE _____ PERIOD _____

Test, Form 1B

SCORE _____

Write the letter for the correct answer in the blank at the right of each question.

What is the circumference of each circle? Use 3.14 for π. Round to the nearest tenth if necessary.

1. A. 12.1 mm   C. 28.3 mm
   B. 14.1 mm   D. 56.5 mm

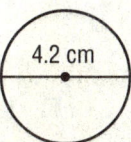

1. _____

2. F. 6.6 cm   H. 13.2 cm
   G. 7.3 cm   I. 26.4 cm

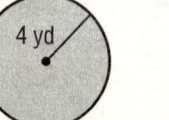

2. _____

What is the area of each circle? Use 3.14 for π. Round to the nearest tenth if necessary.

3. A. 201 yd^2 C. 25.1 yd^2
   B. 50.2 yd^2 D. 12.6 yd^2

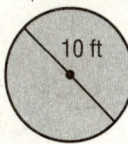

3. _____

4. F. 19.6 ft^2 H. 78.5 ft^2
   G. 31.4 ft^2 I. 314 ft^2

4. _____

5. What is the volume of the rectangular prism?

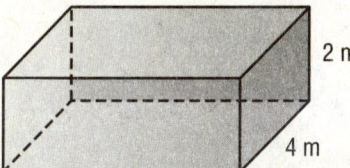

   A. 36 m^3
   B. 62 m^3
   C. 72 m^3
   D. 124 m^3

5. _____

6. A cube has 5-inch edges. What is its volume?
   F. 125 in^3 G. 150 in^3 H. 250 in^3 I. 625 in^3

6. _____

For Exercises 7 and 8, what is the volume of each figure? Round to the nearest tenth if necessary.

7.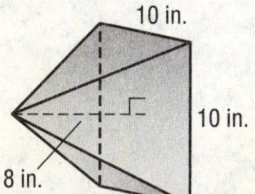

8.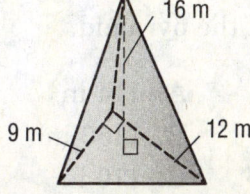

7. _____

A. 800 in^3 C. 320 in^3 F. 192 m^3 H. 576 m^3
B. 433.7 in^3 D. 266.7 in^3 G. 288 m^3 I. 864 m^3

8. _____

Course 2 • Chapter 8 Measure Figures   **183**

Test, Form 1B (continued)

For Exercises 9–11, what is the surface area of each figure? Round to the nearest tenth if necessary.

9.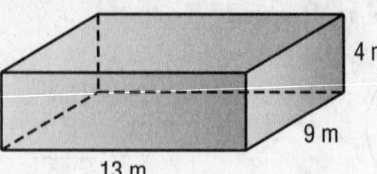

A. 205 m²
B. 234 m²
C. 410 m²
D. 468 m²

9. _____

10.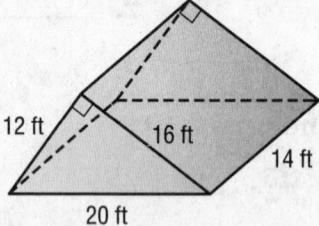

F. 1,344 ft²
G. 896 ft²
H. 864 ft²
I. 672 ft²

10. _____

11.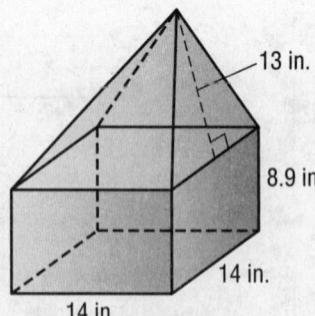

A. 862.4 in²
B. 1,058.4 in²
C. 1,334.2 in²
D. 2,108.4 in²

11. _____

12. Find the area of the composite figure.

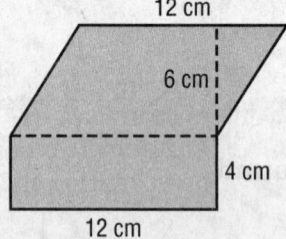

F. 160 cm²
G. 120 cm²
H. 104 cm²
I. 100 cm²

12. _____

13. Find the total surface area of the pyramid.

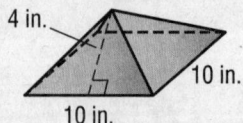

A. 114 in²
B. 1150 in²
C. 170 in²
D. 180 in²

13. _____

NAME _____ DATE _____ PERIOD _____

Test, Form 2A

SCORE _____

Write the letter for the correct answer in the blank at the right of each question.

1. What is the circumference of the circle? Use 3.14 for π. Round to the nearest tenth.
   A. 31.2 yd
   B. 44.1 yd
   C. 88.2 yd
   D. 176.5 yd

   1. _____

2. To the nearest tenth, what is the circumference of a bicycle tire with a radius of 11 inches? Use 3.14 for π.
   F. 34.5 in.   G. 17.3 in.   H. 69.1 in.   I. 95 in.

   2. _____

3. What is the area of the circle? Round to the nearest tenth. Use 3.14 for π.
   A. 1,017.4 mm²
   B. 254.3 mm²
   C. 56.5 mm²
   D. 28.3 mm²

   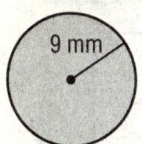

   3. _____

4. A rectangular trunk has a volume of 26,880 cubic inches. The trunk is 4 feet long by 28 inches wide. What is the trunk's height?
   F. 20 in.   G. 60 in.   H. 240 in.   I. 2,880 in.

   4. _____

5. What is the volume of the right triangular prism?
   A. 93.3 m³
   B. 140.3 m³
   C. 280 m³
   D. 560 m³

   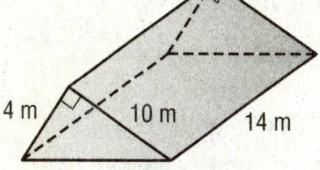

   5. _____

6. What is the volume of the square pyramid?
   F. 384 in³
   G. 192 in³
   H. 132 in³
   I. 128 in³

   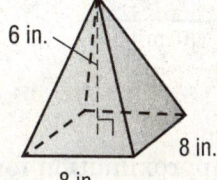

   6. _____

7. What is the surface area of the cube?

   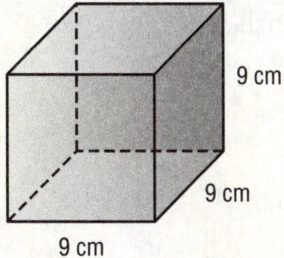

   A. 243 cm²
   B. 364.5 cm²
   C. 486 cm²
   D. 729 cm²

   7. _____

Course 2 • Chapter 8 Measure Figures

185

Test, Form 2A (continued)

8. What is the surface area of the square pyramid?

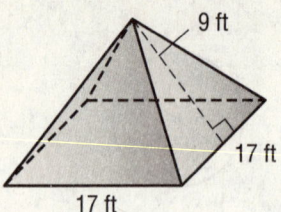

F. 595 ft²   H. 867 ft²
G. 612 ft²   I. 901 ft²

8. _____

9. Find the surface area and volume of the composite figure.

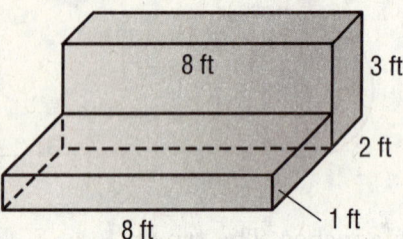

9. _____

10. What is the volume of the triangular pyramid? Round to the nearest tenth if neccessary.

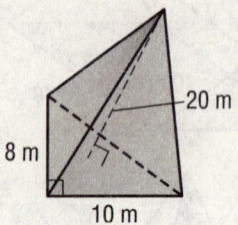

10. _____

11. What is the best approximation for the area of a semicircle with a diameter of 12.5 ft? Use 3.14 for π.

11. _____

12. What is the area of the figure? Round to the nearest tenth if necessary. Use 3.14 for π.

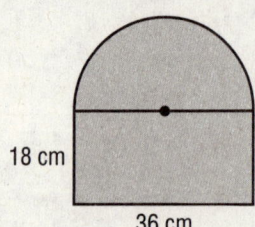

12. _____

NAME _____ DATE _____ PERIOD _____

Test, Form 2B

SCORE _____

Write the letter for the correct answer in the blank at the right of each question.

1. What is the circumference of the circle? Use 3.14 for π. Round to the nearest tenth.
   A. 543 yd   **C.** 41.3 yd
   B. 82.6 yd   **D.** 29.4 yd

   1. _____

2. To the nearest tenth, what is the circumference of a circular pond with a radius of 14 meters? Use 3.14 for π.
   F. 153.9 m   **G.** 87.9 m   **H.** 17.1 m   **I.** 10.1 m

   2. _____

3. What is the area of the circle? Round to the nearest tenth. Use 3.14 for π.
   A. 18.8 ft²   **C.** 39.1 ft²
   B. 28.3 ft²   **D.** 113.0 ft²

   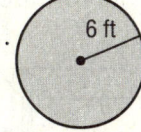

   3. _____

4. A rectangular hat box has a volume of 5,184 cubic inches. The box is 2 feet long by 18 inches wide. What is the hat box's height?
   F. 1 in.   **G.** 12 in.   **H.** 72 in.   **I.** 144 in.

   4. _____

5. What is the volume of the right triangular prism?

   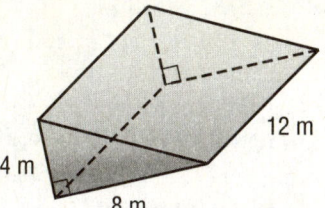

   A. 64 m³
   B. 128 m³
   C. 192 m³
   D. 384 m³

   5. _____

6. What is the volume of the square pyramid?
   F. 64 in³   **H.** 144 in³
   G. 96 in³   **I.** 288 in³

   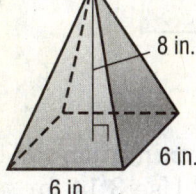

   6. _____

7. What is the surface area of the square pyramid?

   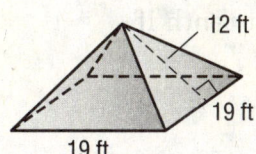

   A. 424 ft²   **C.** 817 ft²
   B. 665 ft²   **D.** 1,444 ft²

   7. _____

Course 2 • Chapter 8 Measure Figures   187

Test, Form 2B (continued)

8. What is the surface area of the cube?

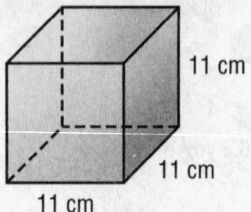

F. 242 cm² H. 726 cm²
G. 484 cm² I. 1,331 cm²

8. _____

9. Find the surface area and volume of the composite figure.

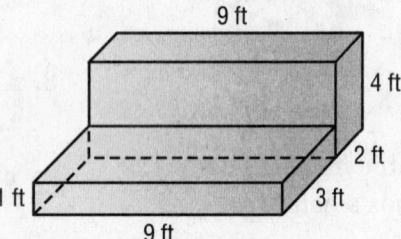

9. _____

10. What is the volume of the triangular pyramid? Round to the nearest tenth if neccessary.

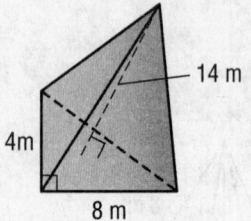

10. _____

11. What is the best approximation for the area of a semicircle with a diameter of 11.8 feet? Use 3.14 for π.

11. _____

12. What is the area of the figure? Round to the nearest tenth if neccessary.

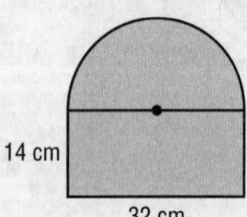

12. _____

188 Course 2 • Chapter 8 Measure Figures

Test, Form 3A

Find the volume of each figure. Round to the nearest tenth if necessary.

1.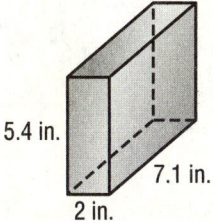

   1. _____

2.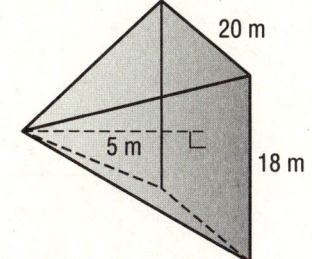

   2. _____

3. A storage shed with a flat roof is 4 yards long by 3 yards wide by $2\frac{1}{2}$ yards tall. A cubic yard is equal to 27 cubic feet. How many cubic feet of storage space does the shed enclose?

   3. _____

4. What is the volume of the square pyramid? Round to the nearest tenth.

   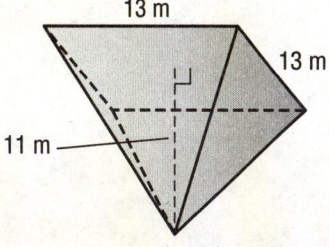

   4. _____

5. What is the circumference of a bagel with a diameter of 11.6 centimeters? Use 3.14 for π. Round to the nearest tenth.

   5. _____

6. Find the area of the circle. Round to the nearest tenth. Use 3.14 for π.

   6. _____

7. Find the area of the shaded region.

   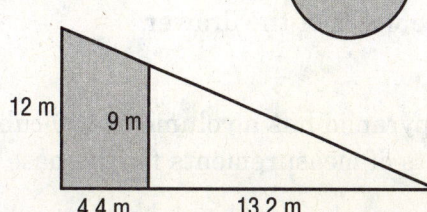

   7. _____

Course 2 • Chapter 8 Measure Figures 189

Test, Form 3A (continued)

8. Find the surface area of the cube.

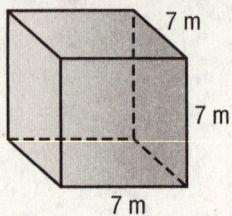

8. _____

9. Find the volume of the pyramid.

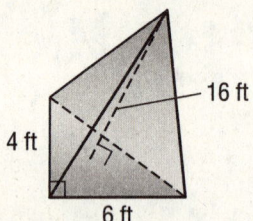

9. _____

10. Find the surface area of the pyramid.

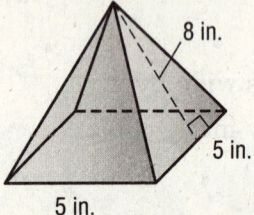

10. _____

11. Find the volume of the composite figure.

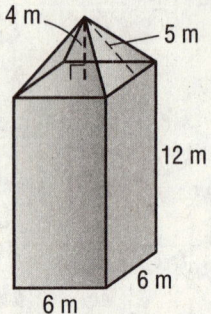

11. _____

12. Find the surface area of the composite figure in Exercise 11.   12. _____

13. A drawer is shaped like a rectangular prism. It has a length of 17 inches and a height of 6 inches. The volume is 1,428 cubic inches. Find the width of the drawer.

13. _____

14. A rectangular pyramid has a volume of 190 cubic centimeters. Find two possible sets of measurements for the base area and height of the pyramid.

14. _____

NAME _____ DATE _____ PERIOD _____

Test, Form 3B

SCORE _____

Find the volume of each figure. Round to the nearest tenth if necessary.

1.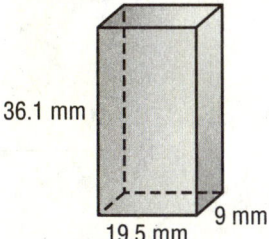

    1. _____

2.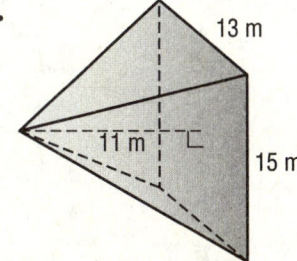

    2. _____

3. A storage shed with a flat roof is 4 yards long by 3 yards wide by $1\frac{1}{2}$ yards tall. A cubic yard is equal to 27 cubic feet. How many cubic feet of storage space does the shed enclose?

    3. _____

4. Find the volume of the square pyramid.

    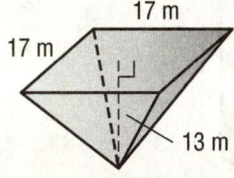

    4. _____

5. What is the circumference of a Ferris wheel with a radius of 22.5 ft? Use 3.14 for π. Round to the nearest tenth.

    5. _____

6. Find the area of the circle. Use 3.14 for π. Round to the nearest tenth.

    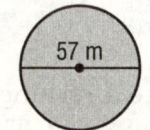

    6. _____

7. Find the area of the shaded region.

    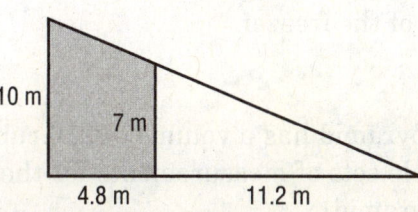

    7. _____

Course 2 • Chapter 8 Measure Figures

191

Test, Form 3B (continued)

8. Find the surface area of the cube.

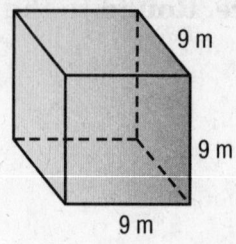

8. _____

9. Find the volume of the pyramid.

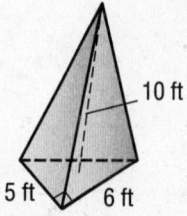

9. _____

10. Find the surface area of the pyramid.

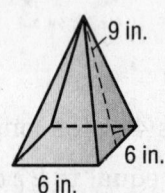

10. _____

11. Find the volume of the composite figure.

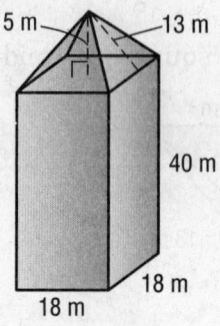

11. _____

12. Find the surface area of the composite figure in Exercise 11.

12. _____

13. A freezer is shaped like a rectangular prism. It has a length of 8 feet and a height of 3 feet. The volume is 54 cubic feet. Find the width of the freezer.

13. _____

14. A rectangular pyramid has a volume of 210 cubic centimeters. Find two possible sets of measurements for the base area and height of the pyramid.

14. _____

NAME _____ DATE _____ PERIOD _____

Are You Ready?

Review

Write $\frac{10}{25}$ in simplest form. Write *simplified* if the fraction is already in simplest form.

$$\frac{10}{25} \overset{\div 5}{\underset{\div 5}{=}} \frac{2}{5}$$ Divide the numerator and denominator by the GCF, 5.

In simplest form, $\frac{10}{25} = \frac{2}{5}$.

Write each fraction in simplest form. Write *simplified* if the fraction is already simplest form.

1. $\frac{5}{45}$

2. $\frac{14}{21}$

3. $\frac{16}{24}$

4. $\frac{6}{18}$

5. $\frac{20}{70}$

6. $\frac{3}{10}$

7. $\frac{6}{16}$

8. $\frac{18}{30}$

9. $\frac{4}{11}$

10. $\frac{21}{30}$

11. $\frac{12}{40}$

12. $\frac{15}{50}$

1. _____
2. _____
3. _____
4. _____
5. _____
6. _____
7. _____
8. _____
9. _____
10. _____
11. _____
12. _____

Course 2 • Chapter 9 Probability

Are You Ready?

Practice

Find each value.

1. 18×7
2. 5×36
3. 42×8
4. 9×23
5. $7 \times 6 \times 5 \times 4$
6. $12 \times 11 \times 10$
7. $9 \times 8 \times 7$
8. $5 \times 4 \times 3 \times 2$
9. $\dfrac{7 \times 6}{3 \times 2}$
10. $\dfrac{5 \times 4 \times 3}{3 \times 2 \times 1}$
11. $\dfrac{10 \times 9 \times 8}{4 \times 3 \times 2}$
12. $\dfrac{8 \times 7 \times 6 \times 5}{7 \times 6 \times 5 \times 4}$

Write each fraction in simplest form. Write *simplified* if the fraction is already in simplest form.

13. $\dfrac{4}{16}$
14. $\dfrac{5}{30}$
15. $\dfrac{7}{11}$
16. $\dfrac{3}{9}$
17. $\dfrac{8}{12}$
18. $\dfrac{2}{10}$

1. _____
2. _____
3. _____
4. _____
5. _____
6. _____
7. _____
8. _____
9. _____
10. _____
11. _____
12. _____

13. _____
14. _____
15. _____
16. _____
17. _____
18. _____

NAME _____ DATE _____ PERIOD _____

Are You Ready?

Apply

Solve.

1. Marjorie ran 3 miles in 21 minutes. Write a fraction, in simplest form, that represents the unit rate in minutes per mile.

2. The average person sleeps 8 hours a day. About how many hours does the average person sleep during a lifetime of 80 years? Use 365 as the number of days in one year.

3. The table shows the number of students in Mrs. Blair's math class that prefer each sport. There are 26 students in the class. What fraction of the class, in simplest form, prefers basketball?

Preferred Sport	
Sport	Students
Baseball	4
Basketball	8
Football	12
Other	2

4. Karen has eight CDs in her music collection. Each CD has nine songs on it. If each song lasts an average of 4 minutes, about how long could Karen play all of her CDs without repeating a song?

5. Hector makes $9 per hour mowing lawns. If he mows lawns for 4 hours each day, how much will he earn after 8 days?

6. A 30-minute television show contained a total of about 6 minutes of commercials. What fraction, in simplest form, of the 30 minutes was devoted to commercials?

7. Leonard baked six batches of chocolate chip cookies. Each batch had 12 cookies. Each cookie had an average of 6 chocolate chips in it. How many chocolate chips were in the six batches of cookies?

8. Jasmine recorded the number of hours she spent doing each activity one day. The table shows the results. What fraction of her day, in simplest form, was spent online?

Daily Activities	
Activity	Hours
Eating	2
Online	4
Reading	3
Sleeping	9
With Friends	6

Course 2 • Chapter 9 Probability

NAME _____ DATE _____ PERIOD _____

Diagnostic Test

Find each value.

1. 22×9

2. 4×17

3. 31×6

4. 7×51

5. $8 \times 7 \times 6 \times 5$

6. $14 \times 13 \times 12$

7. $5 \times 4 \times 3$

8. $11 \times 10 \times 9 \times 8$

9. $\dfrac{9 \times 8}{4 \times 3}$

10. $\dfrac{6 \times 5 \times 4}{3 \times 2 \times 1}$

11. $\dfrac{12 \times 11 \times 10}{5 \times 4 \times 3}$

12. $\dfrac{5 \times 4 \times 3 \times 2}{4 \times 3 \times 2}$

1. _____
2. _____
3. _____
4. _____
5. _____
6. _____
7. _____
8. _____
9. _____
10. _____
11. _____
12. _____

Write each fraction in simplest form. Write *simplified* if the fraction is already in simplest form.

13. $\dfrac{6}{42}$

14. $\dfrac{10}{55}$

15. $\dfrac{14}{28}$

16. $\dfrac{3}{10}$

17. $\dfrac{4}{32}$

18. $\dfrac{18}{27}$

13. _____
14. _____
15. _____
16. _____
17. _____
18. _____

Pretest

Find the total number of outcomes in each situation

1. choosing a number on a number cube and tossing a coin

    1. _____

2. choosing one meat from chicken, beef, or pork; either a baked potato or hash browns; a vegetable of either green beans, carrots, or corn

    2. _____

3. choosing either a basketball or volleyball with color choices of red, white, orange, or blue

    3. _____

A number cube is rolled. Find each probability.

4. P(even number)

    4. _____

5. P(number less than 5)

    5. _____

6. P(2 or 3)

    6. _____

7. P(number greater than 1)

    7. _____

Find the total number of possible outcomes.

8. selecting a president, vice-president, and secretary from a class of 18 students

    8. _____

9. the number of different ways that five people can stand in a line

    9. _____

10. the number of different ways that first and second prizes can be determined from a total of 10 participants, assuming no ties

    10. _____

State whether the events are independent or dependent.

11. tossing a coin and rolling a number cube

    11. _____

12. selecting one marble from a hat, not replacing it, and then selecting a second marble

    12. _____

13. spinning a spinner with four equal sections and tossing a coin

    13. _____

Course 2 • Chapter 9 Probability

NAME _____ DATE _____ PERIOD _____

Chapter Quiz

For Exercises 1 and 2, a number cube labeled one through six is rolled.

1. Find P(2 or 4).

    1. _____

2. Find P(greater than 2).

    2. _____

3. A number cube is rolled. Find the probability of not getting an even number.

    3. _____

4. Brenda can choose between 2 pairs of pants and 3 shirts. How many outfits are possible?

    4. _____

A spinner with four equal-size sections marked M, A, T, and H is spun 100 times. The results are shown below.

Section	Frequency
M	21
A	26
T	25
H	28

5. What is the theoretical probability of landing on A?

    5. _____

6. What is the experimental probability of landing on a vowel?

    6. _____

7. Compare the theoretical probability to the experimental probability of landing on H.

    7. _____

8. Manny rolls a number cube labeled 1 through 6 and spins a spinner with four equal sections, labeled 1, 2, 3, and 4. If both numbers are odd, he wins. Otherwise, Gabriel wins. Use a list to find the sample space. Then find the probability that Gabriel wins.

    8. _____

9. Describe a model that you could use to simulate the outcome of guessing the correct answers to a 25-question true-false test.

    9. _____

198   Course 2 • Chapter 9 Probability

NAME _____ DATE _____ PERIOD _____

Vocabulary Test

complementary events	outcome	simple event
compound event	permutation	simulation
dependent events	probability	theoretical probability
experimental probability fair	random	tree diagram
Fundamental Counting Principle	relative frequency	uniform probability model
independent events	sample space	unfair

Choose from the terms above to complete each sentence.

1. If the outcome of one event does not affect the other event, the events are _____.

    1. _____

2. A _____ is an arrangement, or listing, of objects in which order is important.

    2. _____

3. The _____ can be used to find the total number of possible outcomes by using multiplication.

    3. _____

4. The _____ of rolling a 1 on a number cube is $\frac{1}{6}$.

    4. _____

5. If an event has one outcome or a collection of outcomes, the event is a(an) _____ event.

    5. _____

6. The sum of the probabilities of _____ is 1.

    6. _____

7. The set of all possible outcomes in a probability experiment is the _____.

    7. _____

8. _____, or relative frequency, is based on what actually occurs during an experiment.

    8. _____

9. Organized lists, tables, and _____ can be used to represent sample space.

    9. _____

Define each term in your own words.

10. outcome

    10. _____

11. simulation

    11. _____

Course 2 • Chapter 9 Probability

Standardized Test Practice

Read each question. Then fill in the correct answer on the answer sheet provided by your teacher or on a sheet of paper.

1. Jessica played a game where she spun each of the spinners shown below once. If she spins an even number on Spinner 1, red or yellow on Spinner 2, and a B on Spinner 3, how many possible unique outcomes are there? outcomes are there?

  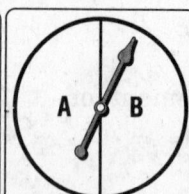

Spinner 1   Spinner 2   Spinner 3

A. 4
B. 8
C. 10
D. 16

2. What is $4 \div \frac{1}{3}$?

F. $\frac{1}{12}$
G. $\frac{4}{3}$
H. 7
I. 12

3. The students in Mrs. Martin's class sell items to raise money for field trips each year. They took a survey to determine which items to sell to other students. The results of the survey are shown in the table. Based on the survey results, what is the probability that a student, selected at random, would buy a drink?

Item	Number of Votes
rings	62
bracelets	27
earrings	21
trading cards	49
snacks	111
small toys	30
drinks	100

A. $\frac{1}{5}$
B. $\frac{1}{4}$
C. $\frac{1}{2}$
D. $\frac{1}{3}$

4. **GRIDDED RESPONSE** Stacy has a spinner and a number cube similar to the ones below. After spinning and rolling the number cube, she will add the two numbers.

What is the probability that the sum of the numbers from the spinner and number cube will be 3 or 4?

5. **GRIDDED RESPONSE** The table shows the total distance traveled by a boat traveling at a constant rate of speed. Based on this information, what will be the distance traveled in miles after 8 hours?

Time (h)	Distance (mi)
2	90
2.5	112.5
3	135
4	180

6. Coach Castillo wanted his team to do a variety of running exercises for practice. To make it more interesting, he used the spinner below to determine which running exercise the team would perform.

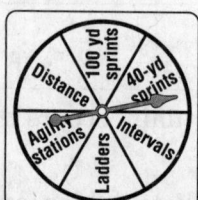

What is the theoretical probability of landing on 40-yard sprints?

F. $\frac{1}{8}$
G. $\frac{1}{6}$
H. $\frac{1}{5}$
I. $\frac{1}{4}$

7. Douglas paid $21 for a pair of jeans at the mall. They were on sale for 20% off. What was the original price before the discount?

   A. $4.20
   B. $5.25
   C. $26.25
   D. $105.00

8. **SHORT RESPONSE** Which point has a coordinate with the greatest absolute value?

   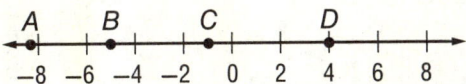

9. A cell phone company charges $35 a month plus $0.30 per text message. Which expression could be used to find the cost for one month of service with b text messages?

   F. $35 + 0.30b$
   G. $35b + 0.30$
   H. $35.30b$
   I. $35b + 0.30b$

10. **SHORT RESPONSE** Corri needs to get milk (M), eggs (E), bread (B), and cereal (C) at the store. Since the bread is close to the cereal, Corri always picks up the cereal right after getting bread.

    List all of the different combinations of ways she can pick up the items she needs. Use the first letter of each item in your list (M, E, B, C).

11. Sierra has 11.5 yards of fabric. She will use 20% of the fabric to make a flag. How many yards of fabric will she use?

    A. 9.2 yd
    B. 8.6 yd
    C. 4.5 yd
    D. 2.3 yd

12. Juan rolled a number cube labeled one through six four times. Each time, the number 3 appeared. If Juan rolls the number cube one more time, what is the probability that 3 will appear?

    F. less than $\frac{1}{6}$
    G. $\frac{1}{6}$
    H. greater than $\frac{1}{6}$
    I. not enough information

13. **EXTENDED RESPONSE** Molly will travel from Trenton to Mayo by car. Suppose she leaves Trenton on one of three routes: 47 North, 129 North, or 26 West, and arrives in Mayo via either 51 North or 27 West. She does not retrace her steps.

    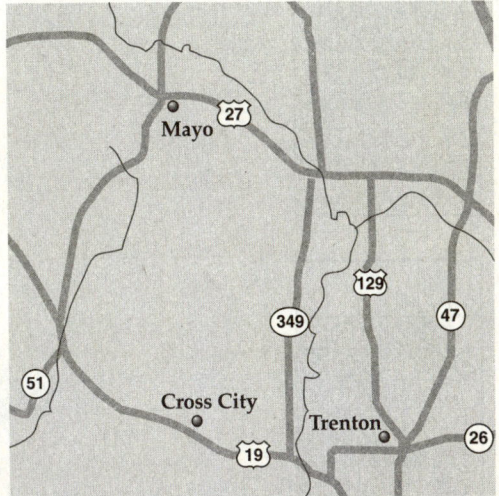

    Part A Based on the map, how many different routes could Molly take for her journey? Create a table, list, or tree diagram to show the possibilities.

    Part B If Molly chooses one route at random, what is the probability she will drive on US 27?

NAME _____ DATE _____ PERIOD _____

Student Recording Sheet

SCORE _____

Use this recording sheet with the Standardized Test Practice pages.

Fill in the correct answer. For gridded-response questions, write your answers in the boxes on the answer grid and fill in the bubbles to match your answers.

1. Ⓐ Ⓑ Ⓒ Ⓓ

2. Ⓕ Ⓖ Ⓗ Ⓘ

3. Ⓐ Ⓑ Ⓒ Ⓓ

4. [grid]

5. [grid]

6. Ⓕ Ⓖ Ⓗ Ⓘ

7. Ⓐ Ⓑ Ⓒ Ⓓ

8. _____

9. Ⓕ Ⓖ Ⓗ Ⓘ

10. _____

11. Ⓐ Ⓑ Ⓒ Ⓓ

12. Ⓕ Ⓖ Ⓗ Ⓘ

Extended Response

Record your answers for Exercise 13 on the back of this paper.

202   Course 2 • Chapter 9 Probability

Extended-Response Test

Demonstrate your knowledge by giving a clear, concise solution to each problem. Be sure to include all relevant drawings and justify your answers. You may show your solutions in more than one way or investigate beyond the requirements of the problem. If necessary, record your answer on another piece of paper.

1. Seven Oaks Middle School is having its Spring fair.

   a. A game uses spinner A below. If a player spins a 1, he or she wins a prize. Explain how to find the theoretical probability of winning a prize.

   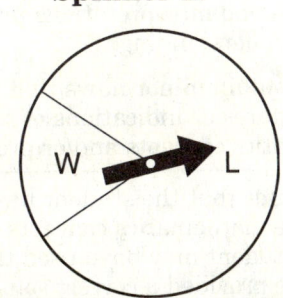

   b. A second game uses spinner B. A player must spin a W to win. Explain how to find the experimental probability of spinning a W.

2. A bag contains 1 white (W), 3 blue (B_1, B_2, B_3), and 2 red (R_1, R_2) marbles.

   a. Use a tree diagram to list all of the possible outcomes for tossing a coin and then drawing a marble from the bag.

   b. Explain what is meant by *independent events*.

   c. Find the probability of tossing a head and drawing a red marble. Explain your reasoning.

   d. Find the probability of drawing two blue marbles if the first marble is not replaced. Explain each step.

3. Rayna and her husband have 8 nieces, 9 nephews, 6 cousins, 3 aunts, and 5 uncles.

   a. Explain how you could use the Fundamental Counting Principle to find how many ways Rayna could choose a team to play charades if she wants one of each type of relative on the team.

   b. How many ways could just the nieces stand in line to limbo? Explain your method.

   c. How many ways could Rayna choose a team to play cards if she wants one of each niece, nephew, or cousin on the team?

Course 2 • Chapter 9 Probability   203

NAME _____ DATE _____ PERIOD _____

Extended-Response Rubric

SCORE _____

Score	Description
4	A score of four is a response in which the student demonstrates a thorough understanding of the mathematics concepts and/or procedures embodied in the task. The student has responded correctly to the task, used mathematically sound procedures, and provided clear and complete explanations and interpretations. The response may contain minor flaws that do not detract from the demonstration of a thorough understanding.
3	A score of three is a response in which the student demonstrates an understanding of the mathematics concepts and/or procedures embodied in the task. The student's response to the task is essentially correct with the mathematical procedures used and the explanations and interpretations provided demonstrating an essential but less than thorough understanding. The response may contain minor flaws that reflect inattentive execution of mathematical procedures or indications of some misunderstanding of the underlying mathematics concepts and/or procedures.
2	A score of two indicates that the student has demonstrated only a partial understanding of the mathematics concepts and/or procedures embodied in the task. Although the student may have used the correct approach to obtaining a solution or may have provided a correct solution, the student's work lacks an essential understanding of the underlying mathematical concepts. The response contains errors related to misunderstanding important aspects of the task, misuse of mathematical procedures, or faulty interpretations of results.
1	A score of one indicates that the student has demonstrated a very limited understanding of the mathematics concepts and/or procedures embodied in the task. The student's response is incomplete and exhibits many flaws. Although the student's response has addressed some of the conditions of the task, the student reached an inadequate conclusion and/or provided reasoning that was faulty or incomplete. The response exhibits many flaws or may be incomplete.
0	A score of zero indicates that the student has provided no response at all, or a completely incorrect or uninterpretable response, or demonstrated insufficient understanding of the mathematics concepts and/or procedures embodied in the task. For example, a student may provide some work that is mathematically correct, but the work does not demonstrate even a rudimentary understanding of the primary focus of the task.

NAME _____ DATE _____ PERIOD _____

Test, Form 1A

SCORE _____

Write the letter for the correct answer in the blank at the right of each question.

For Exercises 1–3, use the spinner at the right. What is each probability written as a fraction in simplest form?

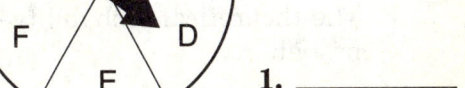

1. $P(C)$
   A. $\frac{1}{8}$
   B. $\frac{1}{6}$
   C. about 50%
   D. 6

   1. _____

2. $P(\text{vowel})$
   F. $\frac{1}{6}$
   G. about 33%
   H. $\frac{1}{2}$
   I. 3

   2. _____

3. $P(\text{not } D)$
   A. 5
   B. $\frac{5}{6}$
   C. about 0.63
   D. $\frac{1}{6}$

   3. _____

For Exercises 4–6, what is the total number of outcomes in each sample space?

4. picking a month of the year and tossing a coin
   F. 2   G. 12   H. 14   I. 24

   4. _____

5. rolling a number cube and tossing a nickel
   A. 2   B. 6   C. 8   D. 12

   5. _____

6. choosing a setting on a washing machine from regular, delicate, or extra dirty; hot, warm, or cold water; regular rinse or extra rinse
   F. 8   G. 9   H. 18   I. 27

   6. _____

7. What is the total number of outcomes for choosing a number from 1 to 10 and a day of the week? Use the Fundamental Counting Principle.
   A. 10   B. 17   C. 63   D. 70

   7. _____

8. A store is handing out coupons worth 10%, 15%, 20%, or 25% off. Each coupon is equally likely to be handed out. Which of the following models could be used to simulate this situation?
   F. flipping a coin four times
   G. spinning a spinner with four equal sections
   H. rolling a number cube labeled one through six one time
   I. rolling a number cube labeled one through six four times

   8. _____

Course 2 • Chapter 9 Probability

205

Test, Form 1A (continued)

For Exercises 9 and 10, Bailey tossed a coin 10 times. The results were 7 heads and 3 tails.

9. What is the experimental probability of tossing tails?
   A. $\frac{1}{3}$ B. $\frac{3}{10}$ C. $\frac{3}{7}$ D. $\frac{1}{2}$

   9. _____

10. What is the best comparison between the theoretical and experimental probability of tossing heads?
    F. The theoretical probability is greater than the experimental probability.
    G. The theoretical probability is less than the experimental probability.
    H. The theoretical probability is equal to the experimental probability.
    I. The theoretical probability is not related to the experimental probability.

    10. _____

11. A bag contains 4 red marbles and 2 white marbles. A marble is selected, kept out of the bag, and then another marble is selected. What is P(red, then white)?
    A. $\frac{4}{25}$ B. $\frac{2}{9}$ C. $\frac{4}{15}$ D. $\frac{1}{3}$

    11. _____

Find each value.

12. $P(8, 3)$
    F. 6   G. 24   H. 336   I. 512

    12. _____

13. $P(10, 4)$
    A. 14   B. 40   C. 5,040   D. 10,000

    13. _____

14. $P(12, 3)$
    F. 15   G. 36   H. 360   I. 1,320

    14. _____

A number cube labeled one though six is rolled and a letter is selected from the word MUSIC. Find each probability.

15. P(2 and S)
    A. $\frac{1}{5}$ B. $\frac{1}{6}$ C. $\frac{1}{11}$ D. $\frac{1}{30}$

    15. _____

16. P(6 and consonant)
    F. $\frac{1}{10}$ G. $\frac{1}{6}$ H. $\frac{3}{5}$ I. $\frac{1}{30}$

    16. _____

17. A jar contains 5 blue marbles, 6 yellow marbles, and 4 green marbles. What is the probability of randomly choosing a yellow marble, not replacing it, and then choosing a blue marble?
    A. $\frac{2}{5}$ B. $\frac{5}{14}$ C. $\frac{1}{7}$ D. $\frac{2}{7}$

    17. _____

NAME _____ DATE _____ PERIOD _____

Test, Form 1B

SCORE _____

Write the letter for the correct answer in the blank at the right of each question.

For Exercises 1–3, use the spinner at the right. What is each probability written as a fraction in simplest form?

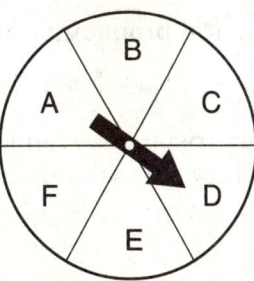

1. $P(D)$
   A. $\frac{1}{6}$
   B. $\frac{4}{5}$
   C. about 50%
   D. 6

   1. _____

2. $P(\text{consonant})$
   F. $\frac{1}{6}$
   G. $\frac{2}{3}$
   H. 0.75
   I. 2

   2. _____

3. $P(\text{not A})$
   A. 0.2
   B. $\frac{1}{6}$
   C. $\frac{5}{8}$
   D. $\frac{5}{6}$

   3. _____

For Exercises 4–6, what is the total number of outcomes in each sample space?

4. tossing a coin and spinning a spinner with five equal sections, 1–5
   F. 2
   H. 10
   G. 7
   I. 15

   4. _____

5. rolling two number cubes
   A. 8
   C. 18
   B. 12
   D. 36

   5. _____

6. choosing water, iced tea, or lemonade; with lemon or lime twist; served in a small or large glass
   F. 7
   H. 8
   G. 10
   I. 12

   6. _____

7. There are seven clarinet players in the concert band. In how many ways can they be seated in seven chairs at a concert? Use the Fundamental Counting Principle.
   A. 5,040
   C. 840
   B. 2,520
   D. 210

   7. _____

8. A store is handing out coupons worth 30%, 35%, or 40% off. Each coupon is equally likely to be handed out. Which of the following models could be used to simulate this situation?
   F. flipping a coin
   G. rolling a number cube labeled one through six three times
   H. spinning a spinner with four equal sections
   I. spinning a spinner with three equal sections

   8. _____

Course 2 • Chapter 9 Probability

207

Test, Form 1B (continued)

For Exercises 9 and 10, Vernon tossed a coin 20 times. The results were 8 heads and 12 tails.

9. What is the experimental probability of tossing heads?
   A. $\frac{1}{8}$ B. $\frac{2}{5}$ C. $\frac{1}{2}$ D. $\frac{3}{5}$

   9. _____

10. What is the best comparison between the theoretical and experimental probability of tossing heads?
    F. The theoretical probability is greater than the experimental probability.
    G. The theoretical probability is less than the experimental probability.
    H. The theoretical probability is equal to the experimental probability.
    I. The theoretical probability is not related to the experimental probability.

    10. _____

11. A bag contains 2 red checkers and 6 black checkers. A checker is selected, kept out of the bag, and then another checker is selected. What is P(black, then red)?
    A. $\frac{1}{9}$ B. $\frac{3}{16}$ C. $\frac{3}{14}$ D. $\frac{9}{16}$

    11. _____

Find each value.

12. $P(7, 2)$
    F. 14   G. 42   H. 49   I. 56

    12. _____

13. $P(9, 3)$
    A. 12   B. 27   C. 72   D. 504

    13. _____

14. $P(5, 4)$
    F. 9   G. 20   H. 120   I. 625

    14. _____

A number cube is rolled and a letter is selected from the word GIRAFFE. Find each probability.

15. P(4 and F)
    A. $\frac{1}{21}$ B. $\frac{1}{6}$ C. $\frac{2}{7}$ D. $\frac{1}{42}$

    15. _____

16. P(even number and vowel)
    F. $\frac{1}{2}$ G. $\frac{3}{7}$ H. $\frac{3}{14}$ I. $\frac{1}{14}$

    16. _____

17. A jar contains 5 blue marbles, 7 yellow marbles, and 8 green marbles. What is the probability of randomly choosing a blue marble, not replacing it, and then choosing a green marble?
    A. $\frac{1}{4}$ B. $\frac{2}{19}$ C. $\frac{8}{19}$ D. $\frac{1}{10}$

    17. _____

NAME _____ DATE _____ PERIOD _____

Test, Form 2A

SCORE _____

Write the letter for the correct answer in the blank at the right of each question.

For Exercises 1 and 2, what would be the total number of outcomes in each sample space?

1. choosing coffee or tea; with cream, milk, or honey; served in a glass or a plastic cup
   A. 6   B. 7   C. 12   D. 24

   1. _____

2. picking a number from 1 to 20 and a letter from the alphabet
   F. 520   G. 260   H. 46   I. 4

   2. _____

3. What is $P(6, \text{ then } 6)$ when spinning the spinner shown at the right twice?
   A. $\frac{1}{49}$ C. $\frac{2}{7}$
   B. $\frac{1}{21}$ D. $\frac{1}{10}$

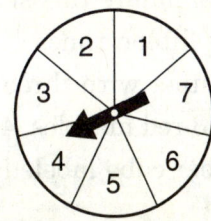

   3. _____

4. Dana has nine pairs of shoes. She wants to organize them in her closet so that they are placed in a row. In how many different ways can Dana arrange nine pairs of shoes in a row? The pairs will not be separated.
   F. 362,880   G. 45,360   H. 6,480   I. 1,080

   4. _____

5. Coach Fraser will select a captain and a co-captain from the students in her physical education class. If there are 22 students from which to select, how many different outcomes are possible?
   A. 484   B. 462   C. 44   D. 22

   5. _____

6. Martin has four books. In how many ways can he arrange them on his bookshelf?
   F. 27   G. 24   H. 12   I. 4

   6. _____

7. Renee tossed 12 heads when tossing a coin 18 times. What is the experimental probability of tossing heads?
   A. about 0.33   B. $\frac{2}{5}$ C. $\frac{2}{3}$ D. 60%

   7. _____

Course 2 • Chapter 9 Probability

NAME _____ DATE _____ PERIOD _____

Test, Form 2A (continued)   SCORE _____

For Exercises 8 and 9, a bag contains 1 red, 2 blue, 4 orange, and 3 purple marbles. A marble is drawn and not replaced. Then a second marble is drawn.

8. What is P(purple, then purple)?
   F. $\frac{1}{15}$ G. $\frac{1}{9}$ H. about 33%   I. 0.09

   8. _____

9. What is P(red, then orange)?
   A. 0.04   B. $\frac{2}{45}$ C. $\frac{4}{81}$ D. about 21%

   9. _____

10. A sports bag contains 3 tennis balls, 4 baseballs, and 8 golf balls. Each is equally likely to be chosen. Which of the following models could be used to simulate this situation?
    F. flipping a coin fifteen times
    G. spinning a spinner with three equal sections
    H. choosing from 3 red marbles, 4 yellow marbles, and 8 blue marbles
    I. rolling a number cube labeled one through six fifteen times

    10. _____

Use the spinner for Exercises 11 and 12.

11. Find P(A or C).

    11. _____

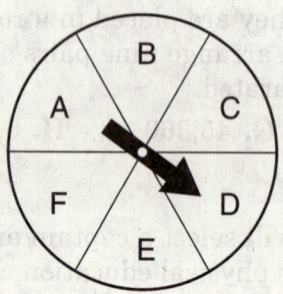

12. Find P(not a vowel).

    12. _____

13. How many ways can Julie, Ayat, Eirene and Bobby finish a race in first, second, third, and fourth place?

    13. _____

14. Melody randomly selected two apples without replacing the first apple from a crate containing 10 Granny Smith apples, 14 Red Delicious apples, 4 Golden Delicious apples, and 18 Braeburn apples. What is the probability that Melody selected a Golden Delicious apple first and a Granny Smith apple second?

    14. _____

15. There are 12 students on the basketball team. In how many ways can the coach set up the starting lineup of 5 players?

    15. _____

210 Course 2 • Chapter 9 Probability

NAME _____ DATE _____ PERIOD _____

Test, Form 2B

SCORE _____

Write the letter for the correct answer in the blank at the right of each question.

For Exercises 1 and 2, what would be the total number of outcomes in each sample space?

1. choosing water, milk, juice, or tea; with or without ice; served in a glass or a plastic cup

   A. 8   **B.** 10   **C.** 16   **D.** 18

   1. _____

2. picking a number from 1 to 30 and a letter from the alphabet

   F. 56   **G.** 150   **H.** 390   **I.** 780

   2. _____

3. What is $P(2, \text{then } 2)$ when spinning the spinner shown at the right twice?

   A. $\frac{1}{4}$ **C.** $\frac{1}{56}$

   B. $\frac{1}{16}$ **D.** $\frac{1}{64}$

   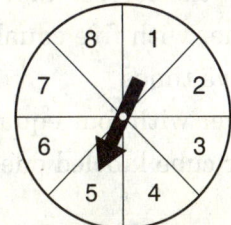

   3. _____

4. Mrs. Blair will select a president and a vice-president from a committee of 8 members. In how many different ways can a president and a vice-president be chosen from the committee?

   F. 1,680   **G.** 336   **H.** 56   **I.** 15

   4. _____

5. Geoff needs to create a password for his E-mail account. The password must have three letters. How many different passwords can he make? Assume he may not use each letter more than once.

   A. 51   **B.** 676   **C.** 15,600   **D.** 358,800

   5. _____

6. In how many ways can five models line up to have their photograph taken?

   F. 120   **G.** 60   **H.** 25   **I.** 20

   6. _____

7. Sienna tossed 10 heads when tossing a coin 24 times. What is the experimental probability of tossing heads?

   A. 0.1   **B.** $\frac{5}{12}$ **C.** 50%   **D.** $\frac{5}{7}$

   7. _____

Course 2 • Chapter 9 Probability

211

Test, Form 2B (continued)

For Exercises 8 and 9, a box contains 1 green, 2 red, 3 pink, and 4 yellow paper clips. A paper clip is drawn and not replaced. Then a second paper clip is drawn.

8. What is P(red, then red)?
   F. $\frac{1}{45}$ G. $\frac{2}{45}$ H. $\frac{1}{25}$ I. 25%

   8. _____

9. What is P(green, then pink)?
   A. 3%   B. $\frac{1}{30}$ C. $\frac{3}{10}$ D. 0.4

   9. _____

10. The questions on a multiple-choice test each have 5 answer choices. Which of the following models could you use to simulate the outcome of guessing the correct answer?
    F. spinning a spinner with five equal sections
    G. flipping a coin five times
    H. spinning a spinner with four equal sections
    I. rolling a number cube labeled one through six five times

    10. _____

Use the spinner for Exercises 11 and 12.

11. Find P(vowel).

    11. _____

12. Find P(not D).

    12. _____

13. How many ways can Gary, Cayla, and Pinak finish a bicycling race in first, second, and third place?

    13. _____

14. Antonio has a CD-player that holds six CDs. He puts six different CDs in the player and the CD player randomly plays a song from any of the CDs. What is the probability that the CD player will play the first song from the first CD and the first song from the sixth CD?

    14. _____

15. The yearbook committee is deciding on the order to place 7 club pages. In how many ways can the committee order the pages if the yearbook club comes last?

    15. _____

NAME _____ DATE _____ PERIOD _____

Test, Form 3A

SCORE _____

Write the correct answer in the blank at the right of each question.

1. Keisha's family is planning a trip to Europe. If they want to visit each of the cities listed in the table at the right, in how many different orders can they do so?

City
Athens
Berlin
London
Paris
Rome

   1. _____

2. Employees at a company are given a three-digit employee identification code. If each digit cannot be repeated, how many different codes are possible?

   2. _____

3. There are 26 students in Mr. Everly's social studies class. Mr. Everly will randomly select one student as spokesperson and a second student as an alternate spokesperson for an upcoming presentation. In how many different ways can they be chosen?

   3. _____

4. Drew spun a spinner with 5 equal sections 75 times. Each section of the spinner was a different color. One of the colors was blue. The outcome of "blue" occurred 30 times. Compare the theoretical to the experimental probability of spinning blue.

   4. _____

5. The table at the right shows the voting preferences for registered voters. Describe a model that you could use to simulate the selection of a candidate.

Candidate	Percent of Voters
Sanchez	45
Ledo	30
Carroll	15
Undecided	10

   5. _____

For Exercises 6 and 7, find the total number of outcomes that will be in each sample space.

6. buying bedroom furniture if you can select one each from 7 dressers, 4 beds, 6 lamps, and 9 night tables

   6. _____

7. tossing a dime, a quarter, a penny, a nickel, and rolling a number cube

   7. _____

8. How many ways can 4 friends sit together at the movies in 4 seats?

   8. _____

Course 2 • Chapter 9 Probability

Test, Form 3A (continued)

Use the spinner to find each probability.

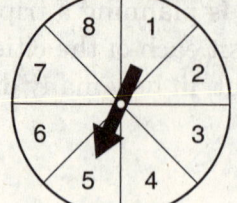

9. P(even number)

9. _____

10. P(2 or 3)

10. _____

11. P(not 4)

11. _____

12. The spinner is spun twice. Find P(5, then 8).

12. _____

A bag contains 4 white beads, 6 red beads, 5 yellow beads, and 5 blue beads. One bead is selected, kept, and another bead is selected.

13. Find P(blue, then blue).

13. _____

14. Find P(white, then red).

14. _____

15. Sohan rolled a number cube 90 times. The outcome of "6" occurred 18 times. Compare the theoretical to the experimental probability of rolling a 6.

15. _____

Find each value.

16. $P(8, 5)$

16. _____

17. $P(10, 2)$

17. _____

18. $P(11, 4)$

18. _____

19. A bowl contains 7 pennies, 9 nickels, and 4 dimes. Elyse removes one coin at random from the bowl and does not replace it. She then removes a second coin at random. What is the probability that both will be dimes?

19. _____

20. There are 100 prize tickets in a bowl, numbered 1 to 100. What is the probability that an even numbered prize ticket will be chosen at random, not replaced, then an odd numbered prize ticket will be chosen? Does this represent an independent or dependent event? Explain.

20. _____

214   Course 2 • Chapter 9 Probability

Test, Form 3B

Write the correct answer in the blank at the right of each question.

1. Derek's family is planning a trip to Asia. If they want to visit each of the cities listed in the table at the right, in how many different orders can they do so?

City
Beijing
Shanghai
Taipei
Tokyo

   1. _____

2. Employees at a company are given a five-digit employee identification code. If each digit cannot be repeated, how many different codes are possible?

   2. _____

3. There are 23 students in Mrs. Sinclair's Spanish class. Mrs. Sinclair will randomly select one student as president and a second student as vice-president. In how many different ways can they be chosen?

   3. _____

4. Adrian spun a spinner with 5 equal sections 85 times. Each section of the spinner was a different color. One of the colors was blue. The outcome of "blue" occurred 20 times. Compare the theoretical to the experimental probability of spinning blue.

   4. _____

5. The table at the right shows the voting preferences for registered voters. Describe a model that you could use to simulate the selection of a candidate.

Candidate	Percent of Voters
Alvarez	20
Jones	40
Mulroney	25
Undecided	15

   5. _____

Exercises 6 and 7, find the total number of outcomes that will be in each sample space.

6. buying bedroom furniture if you can select one each from 8 dressers, 3 beds, 7 lamps, and 4 night tables

   6. _____

7. tossing a dime, a quarter, a penny, and rolling a number cube

   7. _____

8. How many ways can 5 friends sit together at the movies in 5 seats?

   8. _____

Course 2 • Chapter 9 Probability

Test, Form 3B (continued)

Use the spinner to find each probability.

9. P(odd number)

9. _____

10. P(not 3)

10. _____

11. P(4 or 5)

11. _____

12. The spinner is spun twice. Find P(1, then 6).

12. _____

A bag contains 4 white beads, 6 red beads, 5 yellow beads, and 5 blue beads. One bead is selected, kept, and another bead is selected.

13. Find P(red, then red).

13. _____

14. Find P(blue, then yellow).

14. _____

15. Farah rolled a number cube 84 times. The outcome of "2" occurred 12 times. Compare the theoretical to the experimental probability of rolling 2.

15. _____

Find each value.

16. P(4, 4)

16. _____

17. P(6, 3)

17. _____

18. P(9, 5)

18. _____

19. A bowl contains 8 pennies, 7 nickels, and 10 dimes. Elyse removes one coin at random from the bowl and does not replace it. She then removes a second coin at random. What is the probability that both will be nickels?

19. _____

20. There are 26 prize tickets in a bowl, labeled A to Z. What is the probability that a prize ticket with a vowel will be chosen, not replaced, and then another prize ticket with a vowel will be chosen? Does this represent an independent or dependent event? Explain.

20. _____

Are You Ready?

Review

Example 1

SWIMMING The table shows the number of laps swum by sixteen swimmers. What percent of the swimmers swam at least 8 laps?

Number of Laps			
5	2	8	10
6	5	4	8
1	6	7	7
5	4	6	8

Four swimmers swam 8 laps or more.

$\frac{4}{16} = 0.25$ or 25%

So, 25% swam at least 8 laps.

QUIZZES The table shows the results of a math quiz.

Math Quiz Scores				
92	88	95	74	96
84	85	80	91	79
98	80	82	79	94
93	82	83	83	90

1. What were the highest and the lowest scores?

2. What percent of the scores were higher than 84?

BOOKS The circle graph shows the results of a survey.

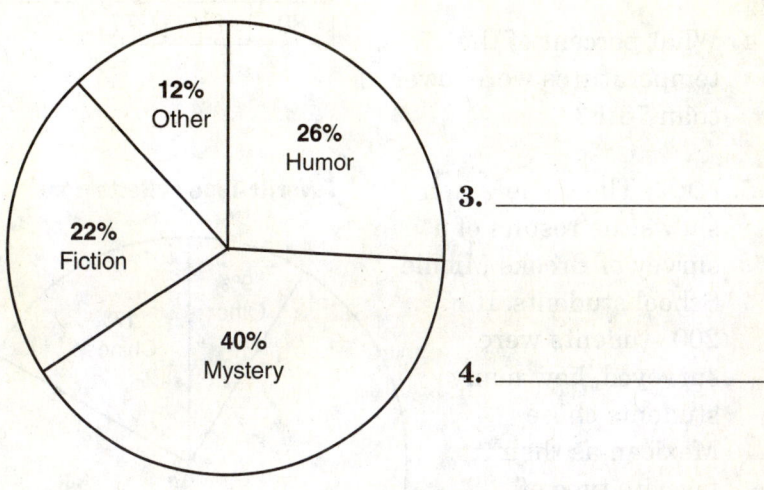

3. If 250 people were surveyed, how many people chose fiction as their favorite type of book?

4. If 160 people were surveyed, how many people chose mystery as their favorite type of book?

5. **SURVEYS** Out of 300 students surveyed, 68% said that pizza was their favorite lunch. How many students surveyed said that pizza was their favorite lunch?

1. _____

2. _____

3. _____

4. _____

5. _____

Course 2 • Chapter 10 Statistics   217

Are You Ready?

Practice

CHORES The bar graph shows the amount of time each person spent doing chores.

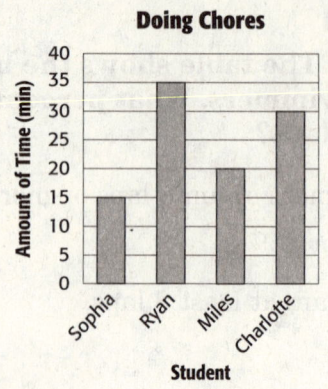

1. Who spent the least amount of time doing chores?

2. Which students spent more than 25 minutes doing chores?

TEMPERATURES The table shows the high temperatures for sixteen days.

High Temperatures (°F)			
79	82	76	80
75	77	81	81
78	83	82	80
80	81	77	79

3. What were the lowest and highest temperatures?

4. What percent of the temperatures were lower than 78°F?

5. **FOOD** The circle graph shows the results of a survey of Brooks Middle School students. If 200 students were surveyed, how many students chose Mexican as their favorite type of restaurant?

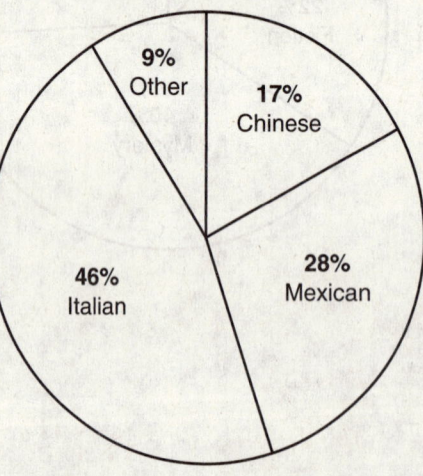

1. _____

2. _____

3. _____

4. _____

5. _____

218 Course 2 • Chapter 10 Statistics

Are You Ready?

Apply

1. **SOCCER** The bar graph shows the number of years that five players have been playing soccer. Which player has been playing soccer the longest?

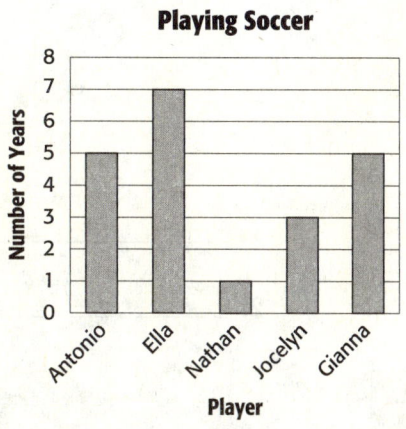

2. **INTERNET** The circle graph shows the results of a survey about the number of days each week that students go online. If 300 students were surveyed, how many students go online 6 or 7 days each week?

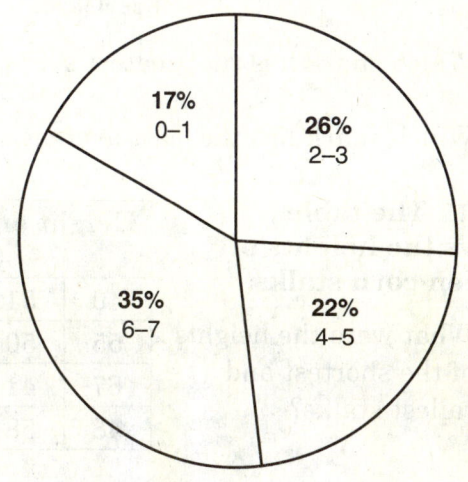

3. **SOCCER** Refer to the bar graph in Exercise 1. Which player(s) have been playing soccer for less than 4 years?

4. **SURVEYS** The Booster Club distributed a survey to 400 families. Of those who received the survey, 76% returned the survey. How many families returned the survey?

5. **MOVIES** In a survey, 220 people were asked to describe the average number of times in a month that they go to a movie theater. Of those surveyed, 45% said they go to the movies more than 3 times a month. How many people surveyed go to the movies more than 3 times a month?

6. **GOLF** The table shows the miniature golf scores of twenty people. What percent of the people had scores of 50 or less?

Miniature Golf Scores				
54	56	50	48	71
63	62	45	51	50
55	65	60	53	70
64	60	55	50	52

Course 2 • Chapter 10 Statistics

Diagnostic Test

SHARKS The bar graph shows the average weight of five different types of sharks.

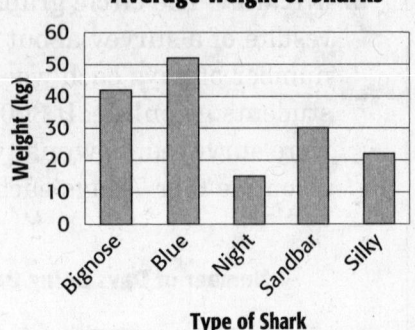

1. Which shark has the greatest average weight? 1. _____

2. Which shark has the least average weight? 2. _____

PLANTS The table shows the heights of sixteen corn stalks.

Height of Corn Stalks (in.)			
40	51	68	57
55	50	57	51
67	41	67	57
48	58	67	59

3. What were the heights of the shortest and tallest stalks? 3. _____

4. What percent of the stalks were taller than 63 inches? 4. _____

5. **FRUIT** The circle graph shows the results of a survey about students' favorite fruit. If 300 students were surveyed, how many students chose bananas as their favorite type of fruit? 5. _____

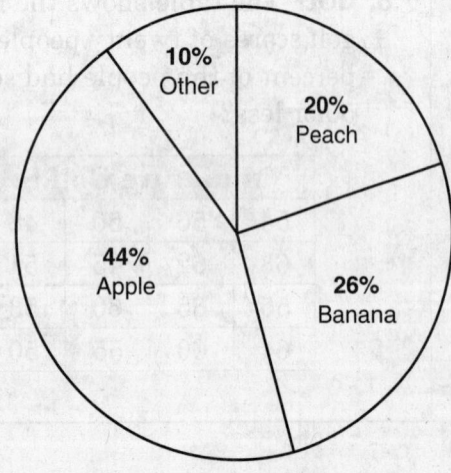

220 Course 2 • Chapter 10 Statistics

NAME _____ DATE _____ PERIOD _____

Pretest

For Exercises 1–4, choose an appropriate type of display for each situation.

1. the grades earned by 20 students on a test

   1. _____

2. the price of a house for the past 20 years

   2. _____

3. the number of families that have dogs, cats, or both as pets

   3. _____

4. the points scored by each person in a basketball game

   4. _____

5. The circle graph shows the results of a class survey. If 30 students were surveyed, how many said they have two pets?

   5. _____

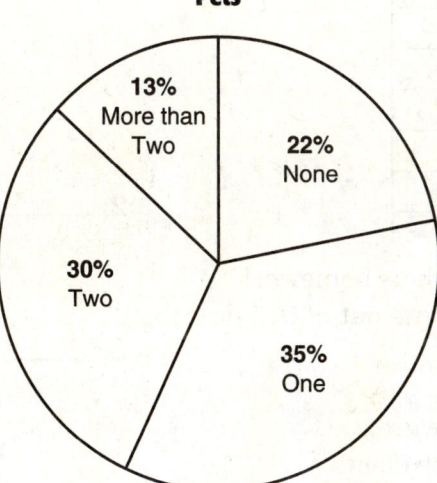

Pets

- 13% More than Two
- 22% None
- 30% Two
- 35% One

6. The graph shows the number of pieces of lost luggage for two different airline companies. Why might this graph be misleading?

   6. _____

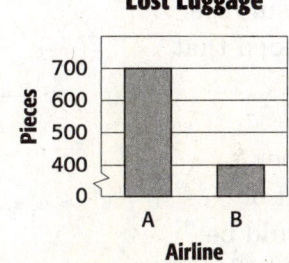

Lost Luggage

Course 2 • Chapter 10 Statistics

NAME _____ DATE _____ PERIOD _____

Chapter Quiz

1. Based on the circle graph, how many more students voted for Raja than for Dayton if 300 students voted?

   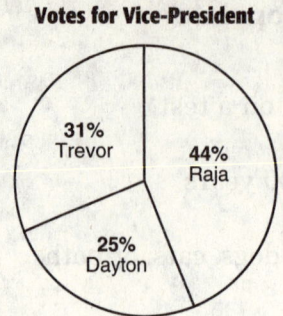
   Votes for Vice-President
   31% Trevor, 44% Raja, 25% Dayton

   1. _____

2. Find the mean, median, and mode of the refrigerator prices shown in the table. Which measurement might be misleading in describing the average cost of a refrigerator?

Refrigerator Prices	
Refrigerator Name	Cost
Quick Freeze	$1,200
Stay Cool	$1,400
Everfrost	$1,200
Cool and Fresh	$400
Thompson Stainless	$1,350

   2. _____

3. A survey found that 78% of students do their homework before 10:00 P.M. Predict how many students out of 975 do their homework before 10:00 P.M.

   3. _____

4. Suppose 11 out of 17 students said they were attending the football game. How many students out of 475 would you expect to attend the football game?

   4. _____

Determine whether each conclusion is valid. Justify your answer.

5. Mr. Dotson wants to know if the neighbors on his street would be interested in a community watch. He surveys every fourth household on the street and concluded that 70% would be interested.

   5. _____

6. Malaya wrote a survey question in the newspaper about changing the school colors. Ninety percent of those who responded in an e-mail said they should be changed. She concludes that 90% of the student body wants the colors changed.

   6. _____

NAME _____ DATE _____ PERIOD _____

Vocabulary Test

biased sample	population	survey
convenience sample	sample	systematic random sample
double box plot	simple random sample	unbiased sample
double dot plot	statistics	voluntary response sample

Choose from the terms above to complete each sentence.

1. A(n) _____ accurately represents the population. 1. _____

2. The _____ is the group being studied. 2. _____

3. A method of collecting data is called a _____. 3. _____

4. A(n) _____ favors one part of the population. 4. _____

5. A _____ is an example of a biased sample because the members of the population chose to be in the sample. 5. _____

6. The branch of mathematics that deals with collecting, organizing, and interpreting data is called _____. 6. _____

7. A(n) _____ consists of two box plots graphed on the same number line. 7. _____

Define each term in your own words.

8. simple random sample 8. _____

9. sample 9. _____

Course 2 • Chapter 10 Statistics

Standardized Test Practice

Read each question. Then fill in the correct answer on the answer sheet provided by your teacher or on a sheet of paper.

1. Which of the following samples would be most representative of the entire student population?
   A. surveying every boy in a gym class
   B. surveying every girl in an art class
   C. surveying every teacher
   D. surveying every 3rd student who enters the school

2. Each spinner is spun once.

   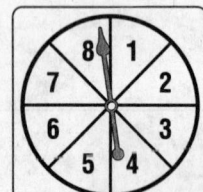 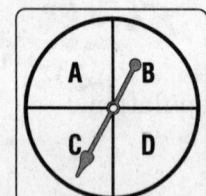

   What is the probability of spinning the number 3 and the letter A?
   F. $\frac{3}{8}$
   G. $\frac{1}{4}$
   H. $\frac{1}{8}$
   I. $\frac{1}{32}$

3. Which of the following equations is equivalent to the equation shown below?

   $$3x + 5 = 7x - 10$$

   A. $10x + 5 = -10$
   B. $4x + 5 = -10$
   C. $7x - 5 = 3x$
   D. $-4x + 5 = -10$

4. **GRIDDED RESPONSE** Neela has 11.5 yards of fabric. She will use 20% of the fabric to make a flag. How many yards of fabric will she use?

5. **GRIDDED RESPONSE** A patio blueprint has a key that shows 1 inch is equal to 12 feet. If the owner wants the length to be 30 feet, how many inches will the length be on the blueprint?

6. The number of ringtones that twelve middle school students have on their cell phones is 14, 8, 7, 6, 5, 5, 10, 11, 8, 8, 6, and 7. Which of the following statements is NOT supported by these data?
   F. Half of the ringtones are below 7.5 and half are above 7.5.
   G. The range of the data is 9 ringtones.
   H. An outlier of the data is 11 ringtones.
   I. About one fourth of the ringtones that the students have are at or above 9.

7. Which box plot represents the data set 8, 12, 21, 15, 20, 9, 16, 14, and 25?

   A.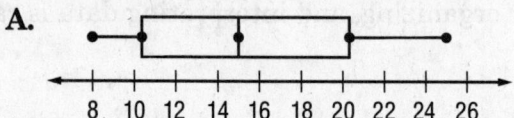

   B.

   C.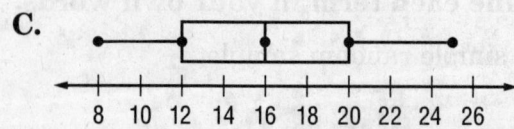

   D.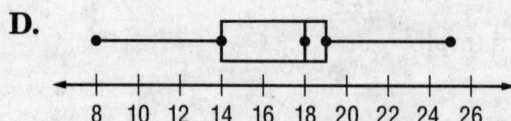

224   Course 2 • Chapter 10 Statistics

NAME _____ DATE _____ PERIOD _____

8. Katherine polled 21 classmates to find out the average number of hours each spends watching television each week. Which of the following displays would be most appropriate to show the individual student responses?

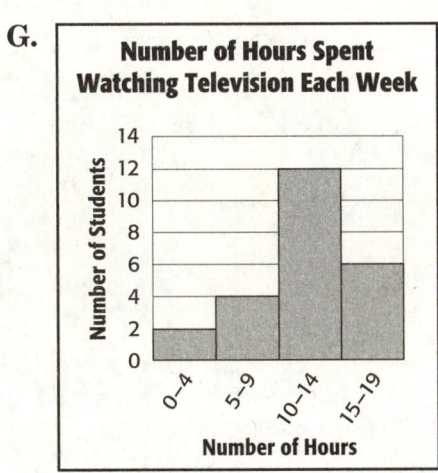

H. Number of Hours Spent Watching Television Each Week

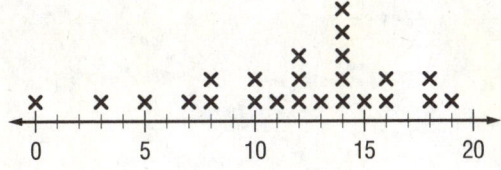

I. Number of Hours Spent Watching Television Each Week

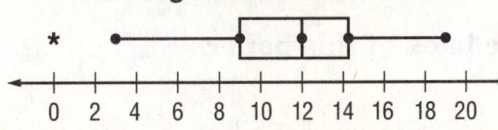

9. The numbers of monthly minutes Gary used on his cell phone for the last eight months are shown below.

Monthly Cell Minutes			
400	550	450	620
550	600	475	425

What is the mode of this data?

A. 550 minutes
B. 450 minutes
C. 475 minutes
D. 400 minutes

10. **SHORT RESPONSE** Mr. Thompson made 20 liters of punch for a party. The punch contained 5 liters of orange juice. Write and solve a proportion to find the percent of orange juice in the punch.

11. **EXTENDED RESPONSE** The table shows how values of a painting increased over ten years.

Year	Value	Year	Value
2005	$350	2010	$1,851
2006	$650	2011	$2,151
2007	$950	2012	$2,451
2008	$1,200	2013	$2,752
2009	$1,551	2014	$3,052

Part A Select and create a display that shows the relationship between years and the value of the painting. Justify your reasoning.

Part B Write a conclusion based on your graph.

Part C Use the graph to predict what the value of the painting will be in 2018.

Course 2 • Chapter 10 Statistics   225

NAME _____ DATE _____ PERIOD _____

Student Recording Sheet

SCORE _____

Use this recording sheet with the Standardized Test Practice pages.

Fill in the correct answer. For gridded-response questions, write your answers in the boxes on the answer grid and fill in the bubbles to match your answers.

1. Ⓐ Ⓑ Ⓒ Ⓓ
2. Ⓕ Ⓖ Ⓗ Ⓘ
3. Ⓐ Ⓑ Ⓒ Ⓓ
4. [grid]
5. [grid]

6. Ⓕ Ⓖ Ⓗ Ⓘ
7. Ⓐ Ⓑ Ⓒ Ⓓ
8. Ⓕ Ⓖ Ⓗ Ⓘ
9. Ⓐ Ⓑ Ⓒ Ⓓ
10. _____

Extended Response

Record your answers for Exercise 11 on the back of this paper.

226   Course 2 • **Chapter 10** Statistics

NAME _____ DATE _____ PERIOD _____

Extended-Response Test

SCORE _____

Demonstrate your knowledge by giving a clear, concise solution to each problem. Be sure to include all relevant drawings and justify your answers. You may show your solution in more than one way or investigate beyond the requirements of the problem. If necessary, record your answer on another piece of paper.

1. David has kept track of his family's grocery bills for the past 10 weeks, as shown in the table.

Week	1	2	3	4	5	6	7	8	9	10
Bill ($)	92	106	129	115	100	84	110	156	98	87

Would you choose to use a histogram, a circle graph, or a line graph to display the data? Explain your choice. Then make a display.

2. Select an appropriate display for the number of athletes compared to the total number of students in junior high school. Explain your choice.

3. **SUNGLASSES** Refer to the line plot below. It shows the prices of sunglasses at a department store.

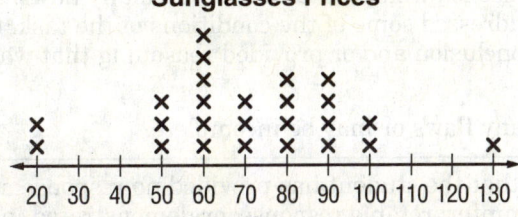

   a. Find the mean, median, and mode.

   b. Which measure best describes the data?

   c. Which measure might be misleading in describing the average price of sunglasses. Explain.

Course 2 • Chapter 10 Statistics

NAME _____ DATE _____ PERIOD _____

Extended-Response Rubric

SCORE _____

Score	Description
4	A score of four is a response in which the student demonstrates a thorough understanding of the mathematics concepts and/or procedures embodied in the task. The student has responded correctly to the task, used mathematically sound procedures, and provided clear and complete explanations and interpretations. The response may contain minor flaws that do not detract from the demonstration of a thorough understanding.
3	A score of three is a response in which the student demonstrates an understanding of the mathematics concepts and/or procedures embodied in the task. The student's response to the task is essentially correct with the mathematical procedures used and the explanations and interpretations provided demonstrating an essential but less than thorough understanding. The response may contain minor flaws that reflect inattentive execution of mathematical procedures or indications of some misunderstanding of the underlying mathematics concepts and/or procedures.
2	A score of two indicates that the student has demonstrated only a partial understanding of the mathematics concepts and/or procedures embodied in the task. Although the student may have used the correct approach to obtaining a solution or may have provided a correct solution, the student's work lacks an essential understanding of the underlying mathematical concepts. The response contains errors related to misunderstanding important aspects of the task, misuse of mathematical procedures, or faulty interpretations of results.
1	A score of one indicates that the student has demonstrated a very limited understanding of the mathematics concepts and/or procedures embodied in the task. The student's response is incomplete and exhibits many flaws. Although the student's response has addressed some of the conditions of the task, the student reached an inadequate conclusion and/or provided reasoning that was faulty or incomplete. The response exhibits many flaws or may be incomplete.
0	A score of zero indicates that the student has provided no response at all, or a completely incorrect or uninterpretable response, or demonstrated insufficient understanding of the mathematics concepts and/or procedures embodied in the task. For example, a student may provide some work that is mathematically correct, but the work does not demonstrate even a rudimentary understanding of the primary focus of the task.

NAME _____ DATE _____ PERIOD _____

Test, Form 1A

SCORE _____

Write the letter for the correct answer in the blank at the right of each question.

1. Antwan wants to know how often the residents in his neighborhood go to the beach. Which sampling method will give valid results?
   A. He asks all the members of the swim team at his school.
   B. He asks all his family members and friends.
   C. He posts a question on a community Web site.
   D. He asks three random households from each street in his neighborhood.

   1. _____

2. Mr. Hou put student names in a hat and selected five names without looking. What type of sample did he form?
   F. simple random sample
   G. systematic random sample
   H. biased sample
   I. convenience sample

   2. _____

3. Seventeen out of 20 teens said they eat breakfast every morning. What is a reasonable prediction for the number of teens out of 1,280 who eat breakfast every morning?
   A. 340   C. 1,088
   B. 940   D. 1,260

   3. _____

4. Which of the following is an appropriate display to show the heights of seventh graders arranged by intervals?
   F. bar graph   H. circle graph
   G. line graph   I. histogram

   4. _____

5. Which of the following is an appropriate display to show the closing price of a stock over the past 3 weeks?
   A. bar graph   C. circle graph
   B. line graph   D. histogram

   5. _____

6. Use the data set $19, $18, $15, $17, $19, $12, $19, and $15. Which measure of center would you use to convince people that the prices are high?
   F. mean   H. mode
   G. median   I. none of these

   6. _____

Course 2 • Chapter 10 Statistics

229

Test, Form 1A (continued)

7. The bar graph shows the heights that students were able to clear on the pole vault. Which statement best tells why the graph could be misleading?

   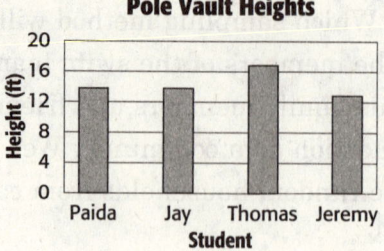

   A. The inverals on the vertical axis are not equal.
   B. The graph title is misleading.
   C. The intervals on the vertical axis make it appear that the heights are nearly the same.
   D. The height should be measured using inches.

   7. _____

8. Katie and Danielle recorded their resting heart rates each morning for ten days. The double dot plot shows their heart rates in beats per minute.

   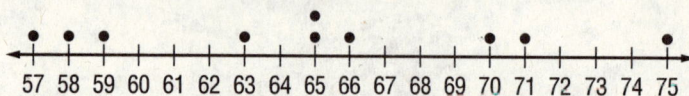

   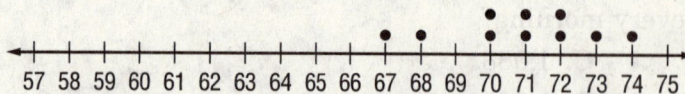

   Which of the following statements is *not* true?
   F. Katie has more varied resting heart rate.
   G. Danielle's heart rates peak between 70 and 72 minutes.
   H. Katie's heart rates peak at 65 minutes.
   I. Katie has a greater average resting heart rate.

   8. _____

9. A survey showed that 70% of students would select roller coasters as their favorite ride at an amusement park. Out of 5,000 students, predict how many would select roller coasters as their favorite ride?
   A. 3,500
   B. 1,500
   C. 350
   D. 150

   9. _____

10. Which measure of center should you use to describe two data sets that are both *not* symmetric?
    F. mean   H. median
    G. mode   I. range

    10. _____

NAME _____ DATE _____ PERIOD _____

Test, Form 1B

SCORE _____

Write the letter for the correct answer in the blank at the right of each question.

1. The manager of a hotel wants to know how often his customers rent boats at a nearby lake. Which sampling method will give valid results?
   A. He asks every tenth customer who checks into the hotel.
   B. He posts a question on the hotel's Web site.
   C. He randomly surveys households in the neighborhood.
   D. He asks every customer in the hotel lobby at noon.

   1. _____

2. To survey a town about traffic concerns, Himani divided the town into eight regions and randomly chose 10 households from each region. What type of sample did she form?
   F. simple random sample
   G. systematic random sample
   H. biased sample
   I. convenience sample

   2. _____

3. Five out of seven teens said they do homework every night. What is a reasonable prediction for the number of teens out of 980 who would do homework every night?
   A. 57   C. 350
   B. 140   D. 700

   3. _____

4. Use the data set $8, $10, $15, $8, $12, $13, $8 and $11. Which measure of center would you use to convince people that the prices are low?
   F. mean   H. mode
   G. median   I. none of these

   4. _____

5. Which of the following is an appropriate display to show the prices of gasoline over the past 3 weeks?
   A. bar graph   C. circle graph
   B. line graph   D. histogram

   5. _____

6. Which of the following is an appropriate display to show the heights of buildings arranged by intervals?
   F. bar graph   H. circle graph
   G. line graph   I. histogram

   6. _____

Course 2 • Chapter 10 Statistics

231

Test, Form 1B (continued)

7. The bar graph compares the number of students that received a grade of an A, B, C or D in Ms. Logan's classroom. Which statement best tells why the graph could be misleading?

   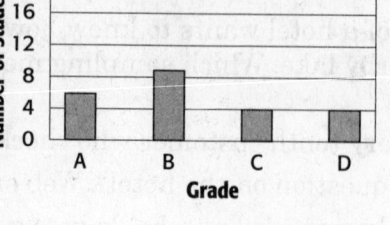

   A. The intervals on the vertical axis are not equal.

   B. The graph title is misleading.

   C. The intervals on the vertical axis make it appear that the number of students that received each grade are nearly the same.

   D. The graph should be a line graph.

   7. _____

8. The double line plot shows the number of students who attended the home games of the baseball team for two recent seasons.

   2009 Student Attendance

   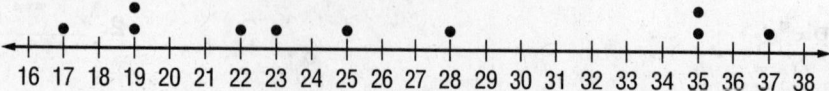

   2010 Student Attendance

   Which of the following statements is *not* true?

   F. The attendance for 2009 was more varied.

   G. The attendance for 2010 was more consistent.

   H. The attendance for 2009 peaked at 23 students.

   I. The attendance for 2010 ranged from 20 to 27.

   8. _____

9. A survey showed that 90% of students would select roller coasters as their favorite ride at an amusement park. Out of 5,000 students, predict how many would select roller coasters as their favorite ride?

   A. 4,500   **C.** 450

   B. 500   **D.** 50

   9. _____

10. Which measure of center should you use to describe two data sets that are both symmetric?

    F. mean   **H.** median

    G. mode   **I.** range

    10. _____

NAME _____ DATE _____ PERIOD _____

Test, Form 2A

SCORE _____

Write the letter for the correct answer in the blank at the right of each question.

1. The double line plot shows the number of hours each month 2 groups of students reported that they watched TV.

Hours of TV for Group 1
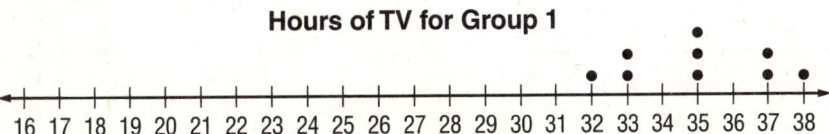

Hours of TV for Group 2
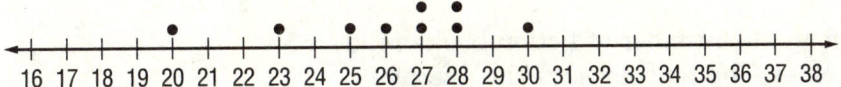

Which of the following statements is true?

A. Group 1 has a greater median number of hours that they watched television. Group 1 has a smaller interquartile range, so the data is less spread out.

B. The mean for group 2 is larger than the mean for group 1.

C. The median for group 2 is larger than the median for group 1.

D. Both sets of data are symmetric. You should use the mean to compare the measures of center and the mean absolute deviation to compare the variations

1. _____

2. A survey found that 5 out of 6 people in a community visit a dentist on a regular basis. If there are 4,320 people in the community, what is a reasonable prediction for the number of people who would visit a dentist on a regular basis?

   F. 720
   G. 864
   H. 2,880
   I. 3,600

2. _____

3. A survey found that 6 out of 8 students own a pet. If there are 320 students in a school, what is a reasonable prediction for the number of students who own a pet?

   A. 240
   B. 80
   C. 24
   D. 8

3. _____

4. Which type of data display would be best for showing the results of a survey on students' favorite school subject?

   F. line plot
   G. bar graph
   H. stem-and-leaf plot
   I. line graph

4. _____

Course 2 • Chapter 10 Statistics

233

Test, Form 2A (continued)

5. The table shows the number of hours Felisa spent sleeping each night for 12 nights. Which type of data display would be best for displaying the data?

Hours Spent Sleeping			
8	6	7	8
10	8	8	6
6	8	8	7

A. line plot
B. scatter plot
C. circle graph
D. line graph

5. _____

6. The line plot below shows the number of fiction books read by one seventh grade class. The teacher says the average number of books read is 6.5. Explain how this could be misleading.

Number of Fiction Books Read

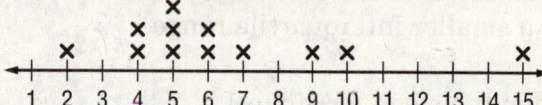

6. _____

7. A basketball coach says that the average height of his starting five basketball players is 72 inches. Explain how this could be misleading. The heights of the five players are shown in the table below.

Players Position	Players Height (in.)
Center	85
Forward	70
Power Forward	69
Shooting Guard	68
Point Guard	68

7. _____

8. To determine what apartment renters want, the manager randomly surveyed eight renters in each of the six buildings in the complex. Out of 48 renters, 64% said they would like more hallway lights. The manager concludes that about two thirds of all the renters would like more hallway lights. Is this conclusion valid? Justify your answer.

8. _____

NAME _____ DATE _____ PERIOD _____

SCORE _____

Test, Form 2B

Write the letter for the correct answer in the blank at the right of each question.

1. The double line plot shows the number of text messages 2 groups of students reported that they sent in one day.

Number of Texts for Group 1

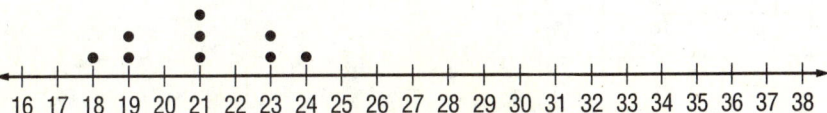

Number of Texts for Group 2

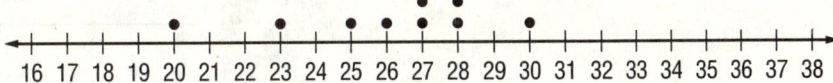

Which of the following statements is true?

A. The mean for group 1 is larger than the mean for group 2.

B. Group 2 has a greater median number of texts that were sent. Group 1 has a smaller interquartile range, so the data is less spread out.

C. The median for group 1 is larger than the median for group 2.

D. Both sets of data are symmetric. You should use the mean to compare the measures of center and the mean absolute deviation to compare the variations

1. _____

2. A survey found that 3 out of 7 people in a community jog on a regular basis. If there are 3,150 people in the community, what is a reasonable prediction for the number of people who would jog regularly?

F. 1,050   H. 1,575
G. 1,350   I. 1,800

2. _____

3. A survey found that 2 out of 8 students do not own a pet. If there are 480 students in a school, what is a reasonable prediction for the number of students who own a pet?

A. 360   C. 36
B. 120   D. 12

3. _____

4. Which type of data display would be best for showing how the height of a plant changes each week during a science experiment?

F. line plot   H. stem-and-leaf plot
G. bar graph   I. line graph

4. _____

Course 2 • Chapter 10 Statistics   235

Test, Form 2B (continued)

5. The table shows the number of hours Felisa spent sleeping each night for 12 nights. Which type of data display would *not* be suitable for displaying the data?

Hours Spent Sleeping			
8	6	7	8
10	8	8	6
6	8	8	7

   A. line plot
   B. bar graph
   C. circle graph
   D. line graph

5. _____

6. The number of toys donated by students in 12 classes is shown below. The principal says the average number of toys donated by each class is 26. Explain how this could be misleading.

   16, 16, 17, 19, 20, 23, 24, 25, 29, 31, 33, 59

6. _____

7. To determine what park visitors like, every tenth visitor is surveyed at the park entrance. Out of 180 visitors, 22% said they would like to have more walking paths. The park manager concludes that about one-fifth of all park visitors would like to have more walking paths. Is this conclusion valid? Justify your answer.

7. _____

8. Which measure of center should you use to describe two data sets that are both symmetric?

8. _____

Course 2 • Chapter 10 Statistics

NAME _____ DATE _____ PERIOD _____

Test, Form 3A

SCORE _____

1. Choose an appropriate type of display. Then make a display.

Price of a Gallon of Regular Gasoline					
Year	2005	2006	2007	2008	2009
Price	$1.36	$1.59	$1.82	$2.34	$2.67

For Exercises 2 and 3, use the box plot that shows student test scores for two classrooms.

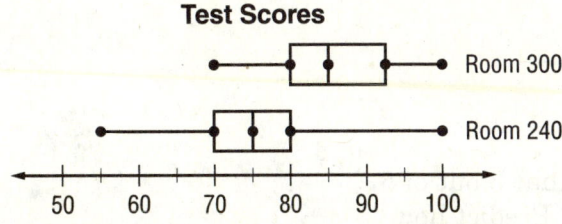

1. _____

2. Which class had a greater range of scores?

2. _____

3. Write an inference you can make about the populations.

3. _____

For Exercises 4 and 5, use the table. It shows the number of books borrowed from a library.

Number of Books			
1	3	15	24
37	55	39	40
35	28	20	0

4. Find the mean, median, and mode of the data.

4. _____

5. If the library wanted to say that it has a high average of books borrowed, which measure of center should they use? Explain.

5. _____

6. The table shows the lengths (in seconds) of student's favorite hit singles. Choose an appropriate display. Then make the display.

Length of Hit Single (s)				
220	150	220	205	256
178	261	258	327	275
166	341	157	208	219
184	265	225	329	248

6. _____

Course 2 • Chapter 10 Statistics

237

NAME _____ DATE _____ PERIOD _____

Test, Form 3A   (continued)   SCORE _____

7. Explain the importance of an unbiased sample.

7. _____

8. Of 350,000 registered voters, 800 were randomly surveyed. Their voting preferences are listed in the table at the right. Predict how many of the registered voters would vote for Sanchez.

Candidate	Percent of Voters
Sanchez	45
Ledo	30
Carroll	15
Undecided	10

8. _____

9. A survey in one middle school showed that 3 out of 5 students enjoy biking on the weekends. Predict how many out of the 485 students in the school would say they enjoy biking on the weekends.

9. _____

10. A pet store owner wants to know what type of pet food her customers prefer. At the bottom of each receipt, she provides a phone number for customers to call in and answer survey questions. Of the 94 responses, she found that 9 prefer organic pet food. She concludes that about 10% of her customers would prefer to feed organic food to their pets. Is this conclusion valid? Justify your answer.

10. _____

11. Jiang Li wants to know if her neighbors want the speed limit on their street reduced. She writes every house number on a piece of paper, puts it into a bag, and selects 15 papers. After surveying these households, she concludes that 75% of the people on her street want the speed limit reduced. Is this conclusion valid? Justify your answer.

11. _____

12. Which measure of variation should you use to describe two data sets that are both symmetric?

12. _____

238   Course 2 • Chapter 10 Statistics

Test, Form 3B

1. Select an appropriate type of display. Then make a display.

High Temperatures on July 4					
Year	2008	2009	2010	2011	2012
°F	82	91	85	77	83

For Exercises 2 and 3, use the box plot that shows students test scores for two classrooms.

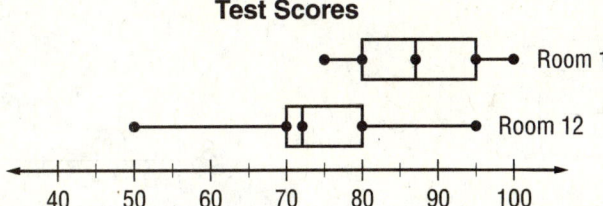

Test Scores

2. Which class had a greater range of scores?

3. Write an inference you can make about the populations.

For Exercises 4 and 5, use the table. It shows the number of students in each extracurricular school club.

Number of Students			
10	12	15	73
13	20	12	16
15	25	9	

4. Find the mean, median, and mode of the data.

5. The principal says the average number of students in a club is 20. How could this be misleading?

6. The table shows the ages of actors who starred in a series of movies. Select an appropriate type of display for the data. Then make a display.

Ages of Actors (yr)					
44	10	24	5	29	30
28	29	18	50	23	3
24	26	8	34	20	24
24	25	9	15	39	16

Course 2 • Chapter 10 Statistics

Test, Form 3B (continued)

7. Name two types of unbiased samples. Then name two types of biased samples.

   7. _____

8. Of 50,000 registered voters, 400 were randomly surveyed. Their voting preferences are listed in the table below. Predict how many of the registered voters would vote for Mulroney.

Candidate	Percent of Voters
Alvarez	20
Jones	40
Mulroney	25
Undecided	15

   8. _____

9. A survey in one middle school showed that 2 out of 9 students help cook meals at home. Predict how many out of the 774 students in the school help cook meals at home.

   9. _____

10. The manager of a recycling center wants to know how many of his customers recycle their batteries. After calling and surveying every fifteenth customer listed in his files, he finds that 46% of them recycle their batteries. He concludes that about half of his customers recycle their batteries. Is this conclusion valid? Justify your answer.

    10. _____

11. Mrs. Melendez wants to know if her neighbors want to hold a neighborhood garage sale. She walks through the neighborhood and asks the people she sees. Because three of the 10 people she saw said *yes*, she concludes that 30% of the people in the city will want to hold a garage sale. Is this conclusion valid? Justify your answer.

    11. _____

12. Which measure of variation should you use to describe two data sets that are both *not* symmetric?

    12. _____

Course 2 Benchmark Test – First Quarter (Chapters 1–2)

1. The table shows the costs of different size jars of peanut butter. Which of the jars has the lowest unit rate?

Comparison Shopping	
Size	Cost
12-oz	$3.00
18-oz	$4.40
25-oz	$6.75
32-oz	$8.25

   A. 12-oz jar

   B. 18-oz jar

   C. 25-oz jar

   D. 32-oz jar

2. The enrollment at a community college this year is 115% of last year's enrollment. If there were 1,240 students enrolled at the college last year, how many students are there this year?

   F. 1,054 students

   G. 1,302 students

   H. 1,378 students

   I. 1,426 students

3. Vicky jogged $2\frac{3}{4}$ miles in $\frac{1}{2}$ hour. What was her average rate of speed in miles per hour?

   A. $1\frac{3}{8}$ miles per hour

   B. $3\frac{1}{4}$ miles per hour

   C. $5\frac{1}{2}$ miles per hour

   D. $6\frac{3}{4}$ miles per hour

4. In a recent survey, 55% of pet owners have more than one pet. If there were 620 pet owners surveyed, which proportion can be used to find the number who own more than one pet?

   F. $\frac{100}{55} = \frac{n}{620}$

   G. $\frac{55}{100} = \frac{n}{620}$

   H. $\frac{55}{100} = \frac{620}{n}$

   I. $\frac{55}{620} = \frac{n}{100}$

5. **SHORT ANSWER** A pair of jeans that normally sells for $35 is on sale for 20% off. Find the sale price of the jeans. Then find the total cost of the jeans if the sales tax rate is 6%.

6. How much simple interest is earned on an investment of $1,250 if the money is invested for 5 years at an annual interest rate of 4.5%?

   A. $1,531.25

   B. $1,306.25

   C. $281.25

   D. $56.25

Course 2 Benchmark Test – First Quarter (continued)

7. SHORT ANSWER Determine whether the relationship between the two quantities in the table is proportional. Explain your reasoning.

Bicycle Rental	
Hours	Cost ($)
0	12.50
1	17.50
2	22.50
3	27.50

8. What is the slope of the line that passes through points R and T?

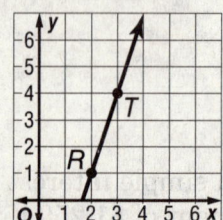

F. $-\frac{3}{1}$

G. $-\frac{1}{3}$

H. $\frac{1}{3}$

I. $\frac{3}{1}$

9. The weight of an object on the moon varies directly as the weight of the object on Earth. A 90-pound object on Earth weighs 15 pounds on the moon. If an object weighs 156 pounds on Earth, how much does it weigh on the moon?

A. 23 pounds

B. 26 pounds

C. 28 pounds

D. 936 pounds

10. A muffin recipe calls for 4 cups of sugar and yields 36 muffins. If Amelia only wants to make 24 muffins, how much sugar will she need?

F. 6 cups

G. $3\frac{3}{4}$ cups

H. $2\frac{2}{3}$ cups

I. $2\frac{1}{2}$ cups

11. A sprinter runs 100 meters in 11.5 seconds. What is the runner's average running rate in meters per second? Round to the nearest tenth.

A. 8.7 meters per second

B. 9.5 meters per second

C. 10.1 meters per second

D. 11.5 meters per second

Course 2 Benchmark Test – First Quarter (continued)

12. Amy earns $7 per hour for babysitting. Which of the following statements is true about the relationship between the number of hours Amy works and her earnings?

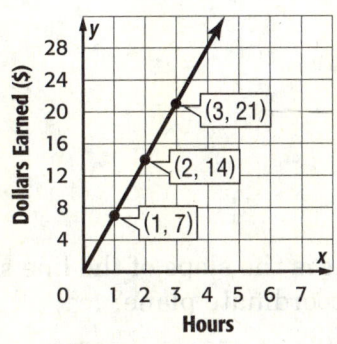
Amy's Earnings

F. The relationship is proportional because the graph of the line passes through the origin and has a constant rate of change.

G. The relationship is proportional because there is not a constant rate of change between the points.

H. The relationship is nonproportional because the points do not form a straight line.

I. The relationship is nonproportional because the line through the points does not intersect the origin.

13. A video game that normally sells for $80 is on sale for $68. What is the percent of discount for the sale price?

A. 18%

B. 17%

C. 15%

D. 12%

14. What is the constant rate of change of the ordered pairs shown in the table?

x	y
2	3
4	7
6	11
8	15

F. 1

G. 2

H. 3

I. 4

15. SHORT ANSWER Estimate 58% of 121 by using 10%. Show your work.

16. Last year there were 43 science projects submitted by students at a science fair. This year there are 52 science projects. To the nearest tenth, what is the percent of change in the number of science projects submitted?

A. 17.3% decrease

B. 17.3% increase

C. 20.9% decrease

D. 20.9% increase

Course 2 • Benchmark Test – First Quarter   243

Course 2 Benchmark Test – First Quarter (continued)

17. Simplify the complex fraction.

$$\frac{\frac{4}{3}}{\frac{2}{5}}$$

F. $\frac{3}{10}$

G. $\frac{8}{15}$

H. $\frac{15}{8}$

I. $\frac{10}{3}$

18. What percent of the figure below is shaded?

A. 45%

B. 40%

C. 20%

D. 18%

19. Mr. Thompson plans to invest $7,500 in a savings account that earns 2.75% simple annual interest. If he makes no other deposits or withdrawals, how much money will Mr. Thompson's account be worth after 10 years?

F. $2,062.50

G. $7,706.25

H. $9,562.50

I. $10,128.25

20. **SHORT ANSWER** Use the percent equation to solve the following problem. Show your work.

98 is 35% of what number?

21. What is the slope of the line shown on the coordinate plane?

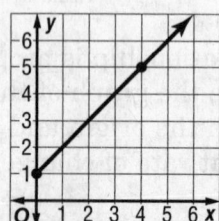

A. 1

B. $\frac{3}{5}$

C. $-\frac{3}{5}$

D. -1

22. Which of the following equations represents a direct variation?

F. $y = x - 1$

G. $y = \frac{x}{3}$

H. $y = x + 5$

I. $y = 2x - 3$

Course 2 Benchmark Test – First Quarter (continued)

23. The bookstore normally sells mechanical pencils for $6.50. This week the pencils are discounted by 25%. To the nearest cent, what is the amount of discount?

    A. $1.30

    B. $1.63

    C. $2.11

    D. $4.88

24. Christy drove 135 miles in 2.5 hours. What was her average speed in miles per hour?

    F. 50 miles per hour

    G. 52 miles per hour

    H. 54 miles per hour

    I. 55 miles per hour

25. **SHORT ANSWER** An electrician charges a $50 fee to make a service call plus $25 per hour he works. Complete the table. Then determine whether the relationship between the two variables is proportional. Explain your reasoning.

Cost of Hiring an Electrician	
Hours	Cost ($)
1	75
2	
3	
4	

Course 2 Benchmark Test – Second Quarter (Chapters 3–5)

1. Which two points represent integers with the same absolute value?

   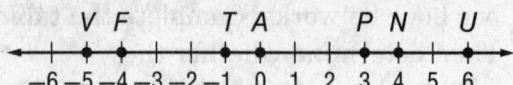

   A. points V and U

   B. points F and P

   C. points T and A

   D. points F and N

2. **SHORT ANSWER** Danielle owes her brother $40. She pays him $25. Write an integer to represent how much she still owes her brother. Explain how you solved.

3. How is the fraction $\frac{19}{30}$ written as a decimal?

   F. 0.63

   G. $0.6\overline{3}$

   H. $0.\overline{63}$

   I. $0.06\overline{3}$

4. Suppose a submarine is diving from the surface of the water at a rate of 80 feet per minute. Which integer represents the depth of the submarine after 7 minutes?

   A. 80

   B. 560

   C. −80

   D. −560

5. What is the simplified form of the algebraic expression shown below?

   $$7w - 6 - 3w + 5$$

   F. $4w - 1$

   G. $w + 2$

   H. $w - 1$

   I. $4w - 6$

6. Which expression is equivalent to the algebraic expression below?

   $$-4(3x - 5)$$

   A. $-x - 5$

   B. $-x - 9$

   C. $-12x + 20$

   D. $-12x - 5$

246   Course 2 • Benchmark Test – Second Quarter

Course 2 Benchmark Test – Second Quarter (continued)

7. Suppose a 24-acre plot of land is being divided into $\frac{1}{3}$-acre lots for a housing development. How many lots will there be in the development?

   F. 8 lots

   G. 27 lots

   H. 56 lots

   I. 72 lots

8. Which property is illustrated by the equation below?

   $$\frac{5}{6} \times \frac{6}{5} = 1$$

   A. Additive Inverse Property

   B. Distributive Property

   C. Associative Property of Multiplication

   D. Multiplicative Inverse Property

9. Which of the following shows the rational numbers in order from least to greatest?

   F. $58\%, 0.6\overline{2}, \frac{31}{50}$

   G. $0.6\overline{2}, \frac{31}{50}, 58\%$

   H. $58\%, \frac{31}{50}, 0.6\overline{2}$

   I. $\frac{31}{50}, 58\%, 0.6\overline{2}$

10. **SHORT ANSWER** Does the pattern below represent an arithmetic sequence? Explain your reasoning.

    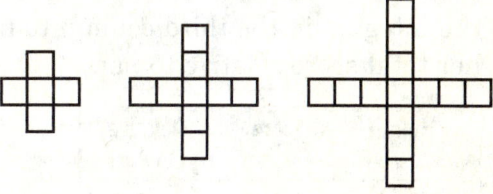

11. Which of the following rational numbers is equivalent to a terminating decimal?

    A. $\frac{17}{20}$

    B. $\frac{17}{22}$

    C. $\frac{17}{24}$

    D. $\frac{17}{26}$

12. Jacob is $5\frac{5}{6}$ feet tall. Linda is $5\frac{1}{4}$ feet tall. How much taller is Jacob?

    F. $\frac{1}{3}$ ft

    G. $\frac{7}{12}$ ft

    H. $\frac{3}{4}$ ft

    I. $1\frac{1}{9}$ ft

Course 2 Benchmark Test – Second Quarter (continued)

13. SHORT ANSWER The table shows Elizabeth's scores for 9 holes of golf. Add the numbers in the middle column to find her total score for 9 holes. Add the integers in the third column to find her total score relative to par.

Hole	Score	Relative to Par
1	4	0
2	5	+1
3	3	0
4	4	0
5	7	+2
6	5	+1
7	4	0
8	5	+1
9	2	−1
Totals	?	?

14. Which of the following linear expressions cannot be factored?

A. $15x - 10$

B. $4x + 8$

C. $3x + 8$

D. $2x - 2$

15. The thickness of a CD is about $\frac{1}{20}$ inch. If Carrie has a stack of 52 CDs, what is the height of the stack?

F. $2\frac{3}{5}$ in.

G. $2\frac{1}{2}$ in.

H. $\frac{5}{13}$ in.

I. $\frac{1}{10}$ in.

16. What is the next number in the pattern?

2,916, −972, 324, −108, 36, …

A. −18

B. −12

C. 12

D. 18

17. Which of the following number sentences represents the model?

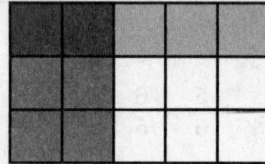

F. $\frac{2}{5} \times \frac{2}{3} = \frac{4}{15}$

G. $\frac{3}{4} \times \frac{1}{3} = \frac{1}{4}$

H. $\frac{2}{3} \times \frac{1}{5} = \frac{2}{15}$

I. $\frac{2}{5} \times \frac{1}{3} = \frac{2}{15}$

18. What is the quotient of the division problem?

$$\frac{-44}{4}$$

A. −11

B. −4

C. 4

D. 11

Course 2 Benchmark Test – Second Quarter *(continued)*

19. Which of the following represents the expression below simplified?

$$(4x - 1) + (-6x + 3)$$

F. $-2x + 2$

G. $-2x + 3$

H. $-2x - 1$

I. $3x - 3$

20. Angela painted $\frac{3}{8}$ of a room. Todd painted $\frac{2}{5}$ of the same room. What part of the room has been painted?

A. $\frac{1}{40}$

B. $\frac{5}{13}$

C. $\frac{31}{40}$

D. $\frac{15}{16}$

21. What is the result when the expression $(6x - 3)$ is subtracted from $(-3x + 2)$?

F. $9x - 5$

G. $-9x + 5$

H. $3x - 1$

I. $-3x + 1$

22. The models below represent the portion of a pizza that Reggie and Edgar have each eaten.

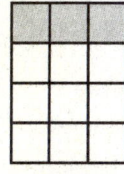 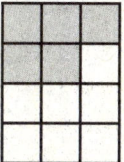

Reggie   Edgar

How much more of the pizza has Edgar eaten than Reggie?

A. $\frac{2}{3}$

B. $\frac{1}{4}$

C. $\frac{1}{6}$

D. $\frac{1}{12}$

23. SHORT ANSWER James is using properties of real numbers to prove that $3(-1) = -3$. Identify the missing properties from his proof.

Statements	Properties
$3(0) = 0$	Multiplicative Property of Zero
$3[(-1) + 1] = 0$	a.
$3(-1) + 3(1) = 0$	b.
$3(-1) + 3 = 0$	c.
$3(-1) = -3$	d.

Course 2 Benchmark Test – Second Quarter (continued)

24. SHORT ANSWER Write the next three terms of the arithmetic sequence below.

1, 9, 17, 25, 33, …

25. Overnight the low temperature dropped to −6 degrees Fahrenheit. If the high temperature during the day was 11 degrees Fahrenheit, what was the difference between the high and low temperatures?

F. 5°F

G. 17°F

H. −5°F

I. −17°F

Course 2 Benchmark Test – Third Quarter (Chapters 6–7)

1. Suppose the length of each side of a square is increased by 5 feet. If the perimeter of the square is now 56 feet, what were the original side lengths of the square?

   A. 9 ft

   B. 11 ft

   C. 14 ft

   D. 36 ft

2. Which operation should be performed first to solve the inequality below?

   $$-3x + 5 \leq 23$$

   F. add 5 to each side

   G. divide each side by −3

   H. reverse the inequality symbol

   I. subtract 5 from each side

3. What is the measure of x in the figure below?

   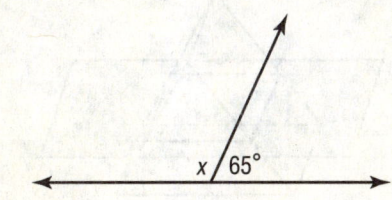

   A. 25°

   B. 65°

   C. 115°

   D. 125°

4. Which of the following shows a straight angle?

   F.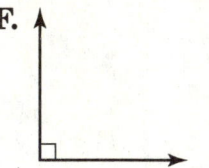

   G.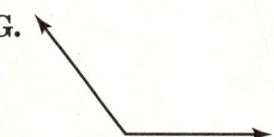

   H.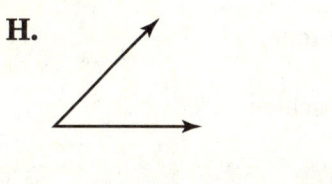

   I. ←——————→

5. What is the solution to the equation below?

   $$\frac{x}{3} = -6$$

   A. −18

   B. −9

   C. −3

   D. −2

6. **SHORT ANSWER** The sum of the measures of the angles of a triangle is 180°. Write and solve an equation to find the missing measure in the figure below. Show your work.

   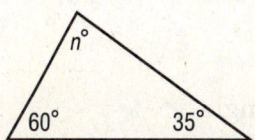

Course 2 Benchmark Test – Third Quarter (continued)

7. Which of the following best classifies the triangle below by its angles and sides?

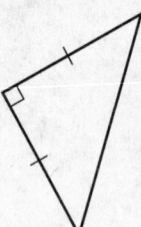

F. acute, isosceles

G. acute, equilateral

H. right, scalene

I. right, isosceles

8. **SHORT ANSWER** A shipping company charges $3.50 plus $0.85 per pound to ship a package. Janet shipped a package and the total charge was $8.60. Write and solve an equation to find the weight of the package.

9. Which of the following describes the shape resulting from the cross section below?

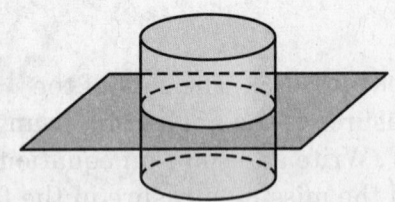

A. circle

B. oval

C. rectangle

D. line

10. What is the solution to the equation below?

$$-\frac{5}{4}x + \frac{2}{5} = -\frac{13}{30}$$

F. $\frac{2}{75}$

G. $\frac{25}{24}$

H. $\frac{2}{3}$

I. $\frac{11}{12}$

11. Terrance is making a scale model of a car that is 16 feet long. He is using the scale 1 inch = 2.5 feet. How long is Terrance's model?

A. 5.8 in.

B. 6.4 in.

C. 28 in.

D. 40 in.

12. Which of the following describes the shape resulting from the cross section below?

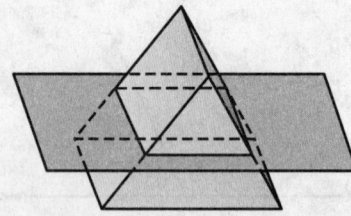

F. rectangle

G. square

H. triangle

I. parallelogram

Course 2 Benchmark Test – Third Quarter (continued)

13. SHORT ANSWER Carla and Mandy are solving the inequality below.

$$-4x \geq 12$$

Carla says the solution is $x \leq -3$, while Mandy says the solution is $x \geq -3$. Which student is correct? What mistake was made by the other student?

14. Which number line shows the solution to the inequality below?

$$v - 2 > 1$$

A. ⟵————◯——⟶
   −5 −4 −3 −2 −1 0 1 2 3 4 5

B. ——◯————————⟶
   −5 −4 −3 −2 −1 0 1 2 3 4 5

C. ————————◯——⟶
   −5 −4 −3 −2 −1 0 1 2 3 4 5

D. ⟵————◯————⟶
   −5 −4 −3 −2 −1 0 1 2 3 4 5

15. Angles R and Z are complementary. If $m\angle R = 26°$, what is the measure of angle Z?

F. 26°

G. 64°

H. 74°

I. 154°

16. Tien bought movie tickets for herself and two of her friends. She paid $8.50 for each ticket. If Tien has $14.50 left, how much money did she have before she bought the movie tickets?

A. $28.00

B. $31.50

C. $37.50

D. $40.00

17. The angle measures of a triangle are 28°, 70°, and 82°. Which of the following best classifies the triangle by its angle measures?

F. acute

G. obtuse

H. right

I. scalene

18. What type of angle is shown below?

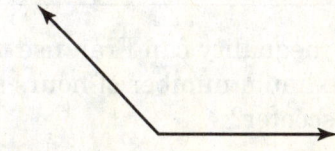

A. acute

B. right

C. obtuse

D. straight

Course 2 Benchmark Test – Third Quarter (continued)

19. Which of the following is a possible cross section of the figure below?

    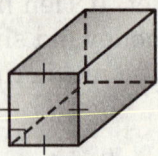

    F. triangle

    G. hexagon

    H. square

    I. trapezoid

20. What is the scale factor of a drawing if the scale is 1 inch = 6 feet?

    A. $\frac{1}{72}$

    B. $\frac{1}{6}$

    C. 6

    D. 72

21. Fran wants to rent a scooter for the afternoon, but she can spend no more than $50.

Scooter Rental
First Hour..............................$12.50
Each Additional Hour..............$7.50

    Which inequality can Fran use to find the maximum number of hours she can rent a scooter?

    F. $12.5 + 7.5n \leq 50$

    G. $12.5 + 7.5n < 50$

    H. $12.5n + 7.5 \leq 50$

    I. $20n < 50$

22. **SHORT ANSWER** Solve the equation below. Check your answer.

    $2(x + 5) = 16$

23. Five more than twice a number is equal to 19. What is the number?

    A. 6

    B. 7

    C. 12

    D. 28

24. Two angle measures in a parallelogram are labeled. Which term best describes the angles?

    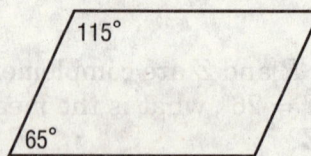

    F. complementary

    G. acute

    H. obtuse

    I. supplementary

Course 2 Benchmark Test – Third Quarter (continued)

25. SHORT ANSWER Jamal built the three-dimensional figure below using blocks.

front

Sketch the front, side, and top views of the figure.

Course 2 Benchmark Test – End of Year

1. If Michelle rollerblades around a circular track with a radius of 80 meters, how far does she skate? Use 3.14 for π. Round to the nearest tenth.

   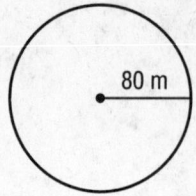

   A. 251.2 m

   B. 502.4 m

   C. 12,352 m

   D. 20,096 m

2. A sprinter runs 400 meters in 54 seconds. What is the runner's average running rate in meters per second? Round to the nearest tenth.

   F. 8.5 meters per second

   G. 7.8 meters per second

   H. 7.4 meters per second

   I. 6.8 meters per second

3. **SHORT ANSWER** Find the slope of the line that passes through points A and B. Show your work.

   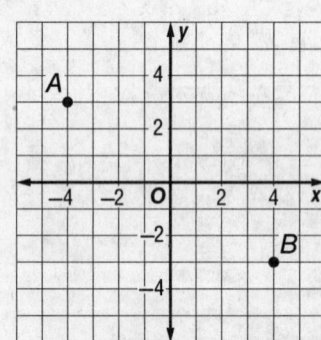

4. The weight of an object on Mars varies directly as the weight of the object on Earth. A 90-pound object on Earth weighs 34 pounds on Mars. If an object weighs 135 pounds on Earth, how much does it weigh on Mars?

   A. 51 pounds

   B. 63 pounds

   C. 219 pounds

   D. 357 pounds

5. A jar contains 3 pennies, 5 nickels, 4 dimes, and 6 quarters. If a coin is selected at random, what is the probability of selecting a penny?

   F. $\frac{5}{18}$

   G. $\frac{2}{9}$

   H. $\frac{1}{3}$

   I. $\frac{1}{6}$

6. Which expression is equivalent to the algebraic expression below?

   $$3(-2x - 1)$$

   A. $x + 2$

   B. $x - 1$

   C. $-6x - 3$

   D. $-6x - 1$

Course 2 Benchmark Test – End of Year (continued)

7. **SHORT ANSWER** A cereal company is giving away 1 of 6 different prizes in each box of cereal. Describe a simulation you could use to estimate the number of boxes needed to get all 6 prizes.

8. Which three-dimensional figure is modeled by the net below?

   F. rectangular prism
   G. triangular prism
   H. square pyramid
   I. rectangular pyramid

9. What is the vertical cross section of a cylinder?

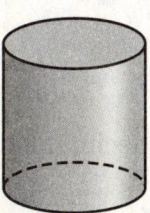

   A. circle
   B. oval
   C. rectangle
   D. point

10. What is the probability of tossing a penny and landing on heads three times in a row?

    F. $\frac{3}{2}$
    G. $\frac{1}{2}$
    H. $\frac{1}{4}$
    I. $\frac{1}{8}$

11. What type of angle is shown below?

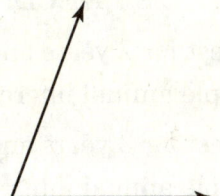

    A. acute
    B. right
    C. obtuse
    D. straight

12. What is the scale factor of a drawing if the scale is 1 inch = 4 feet?

    F. $\frac{1}{48}$
    G. $\frac{1}{4}$
    H. 4
    I. 48

Course 2 Benchmark Test – End of Year (continued)

13. Megan surveyed a random sample of 60 students at her school and found that 42 of them ride the bus to school each day. If there are 320 students at Megan's school, about how many of them ride the bus to school each day?

    A. 348 students

    B. 224 students

    C. 188 students

    D. 132 students

14. **SHORT ANSWER** The advertisement below shows the terms of a certificate of deposit (CD) at a local bank.

    > **Super CD!**
    > ▸ Invest for 2 years and earn 2.75% simple annual interest.
    > ▸ Invest for 3 years and earn 3.25% simple annual interest.
    > ▸ Invest for 4 years and earn 3.75% simple annual interest.
    > See an associate today!

    Suppose Robert invests $1,200 in the CD for a period of 3 years. How much interest will he earn? How much will Robert have after 3 years?

15. Last summer there were 88 players at Coach Rodriguez's basketball camp. This year there are 125% of this number of players. How many players are there at camp this year?

    F. 70 players

    G. 98 players

    H. 106 players

    I. 110 players

16. What is the volume of the pyramid shown below?

    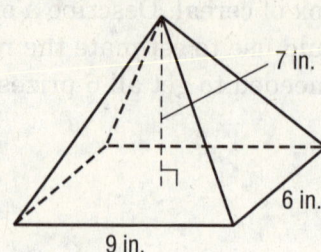

    A. 126 in³

    B. 189 in³

    C. 221 in³

    D. 378 in³

17. What is the constant rate of change of the ordered pairs?

x	y
1	2
3	10
5	18
7	26

    F. 8

    G. 6

    H. 4

    I. 2

18. What is the decimal equivalent of the fraction $\frac{32}{45}$?

    A. 0.71

    B. $0.7\overline{1}$

    C. $0.\overline{71}$

    D. $0.0\overline{71}$

Course 2 Benchmark Test – End of Year (continued)

19. Kyle wants to determine the most popular sport among students at his school. Which of the following will likely result in a biased sample?

    F. surveying every 5th student standing in the lunch line

    G. surveying a random sample of 3 students from each homeroom

    H. surveying a random sample of 25 students attending a school football game

    I. surveying every 10th student who enters the school one morning

20. Last year there were 29 students at a creative writing workshop. This year 35 students attended the workshop. To the nearest tenth, what is the percent of change in the number of students in attendance?

    A. 20.7% decrease

    B. 20.7% increase

    C. 17.1% decrease

    D. 17.1% increase

21. In a recent survey, 88% of shoppers at a grocery store said they would be interested in a rewards program. If there were 450 shoppers surveyed, which proportion can be used to find the number who are interested in a rewards program?

    F. $\frac{100}{88} = \frac{n}{450}$

    G. $\frac{88}{450} = \frac{n}{100}$

    H. $\frac{88}{100} = \frac{450}{n}$

    I. $\frac{88}{100} = \frac{n}{450}$

22. Which property is illustrated by the equation below?

    $$7 + (-7) = 0$$

    A. Additive Inverse Property

    B. Distributive Property

    C. Associative Property of Addition

    D. Additive Identity Property

23. Which of the following shows the rational numbers in order from least to greatest?

    F. $81.5\%, 0.81\overline{5}, \frac{33}{40}$

    G. $81.5\%, \frac{33}{40}, 0.81\overline{5}$

    H. $0.81\overline{5}, \frac{33}{40}, 81.5\%$

    I. $0.81\overline{5}, 81.5\%, \frac{33}{40}$

24. **SHORT ANSWER** The line graph shows the performance of a stock over a 5-day period. Describe what is misleading about the data display.

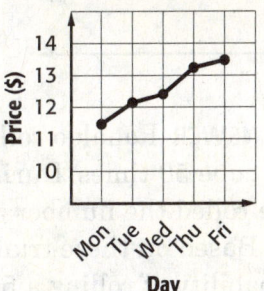

Stock Performance

Course 2 Benchmark Test – End of Year (continued)

25. How many blocks were needed to make the rectangular prism below?

  A. 54 blocks

  B. 72 blocks

  C. 84 blocks

  D. 108 blocks

26. Which of the following angles would be classified as an acute angle?

  F.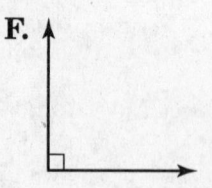

  G.

  H.

  I.

27. **SHORT ANSWER** Ronaldo rolled a number cube 50 times. During these trials he rolled the number 5 a total of 7 times. Based on these trials, what is the probability of rolling a 5? Does this represent a theoretical or experimental probability? Explain.

28. Which of the following linear expressions *cannot* be factored?

  A. $15x + 22$

  B. $12x - 10$

  C. $8x - 2$

  D. $7x + 21$

29. Which of the following number sentences represent the model shown below?

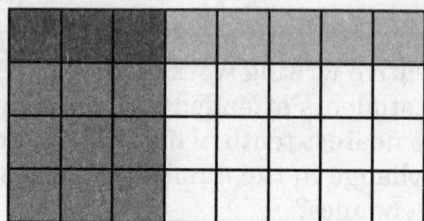

  F. $\frac{3}{4} \times \frac{1}{8} = \frac{3}{32}$

  G. $\frac{3}{8} \times \frac{3}{4} = \frac{9}{32}$

  H. $\frac{3}{8} \times \frac{1}{4} = \frac{3}{32}$

  I. $\frac{1}{4} \times \frac{1}{3} = \frac{1}{12}$

30. Which of the following rational numbers is equivalent to a repeating decimal?

  A. $\frac{24}{60}$

  B. $\frac{30}{64}$

  C. $\frac{29}{50}$

  D. $\frac{35}{60}$

Course 2 Benchmark Test – End of Year (continued)

31. The angle measures of a triangle are 33°, 94°, and 53°. Which of the following best classifies the triangle by its angle measures?

F. acute

G. obtuse

H. right

I. scalene

32. SHORT ANSWER Write and solve an equation to find the missing measure. Show your work.

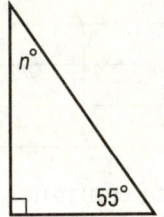

33. What is the measure of x in the figure below?

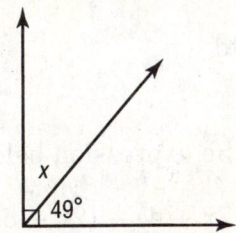

A. 31°

B. 41°

C. 49°

D. 131°

34. A large pizza at Angelo's Pizzeria has a diameter of 14 inches. What is the area of the pizza? Use 3.14 for π. Round to the nearest tenth.

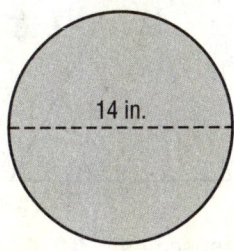

F. 44.0 in²

G. 122.7 in²

H. 153.9 in²

I. 615.4 in²

35. A home improvement store normally sells 20-foot extension ladders for $225. This week the ladders are discounted by 20%. What is the sale price of the ladders?

A. $180

B. $165

C. $60

D. $45

36. SHORT ANSWER A computer store builds custom computers by allowing customers to choose 1 of 4 different CPUs, 1 of 8 hard drives, and 1 of 3 video cards. How many different computers are possible?

Course 2 Benchmark Test – End of Year (continued)

37. Which of the following best classifies the triangle below by its angles and sides?

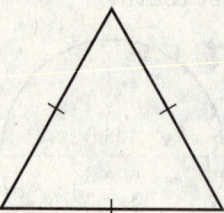

F. acute, isosceles

G. acute, equilateral

H. acute, scalene

I. obtuse, equilateral

38. In an obstacle course race, how many ways can five finalists be ordered?

A. 1

B. 5

C. 24

D. 120

39. SHORT ANSWER Compare and contrast the data represented in the double box plot below.

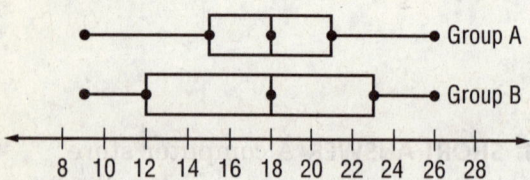

40. What is the solution to the equation below?

$$\frac{7}{8}\left(x - \frac{1}{2}\right) = -\frac{49}{80}$$

F. $-\frac{6}{5}$

G. $-\frac{1}{5}$

H. $\frac{1}{5}$

I. $\frac{6}{5}$

41. The table shows the number of yards jogged by Kaylee each minute.

Time (min)	Distance (yd)
1	175
2	350
3	525
4	700

If the pattern continues, how many yards will Kaylee have jogged after 20 minutes?

A. 875 yd

B. 1,750 yd

C. 3,500 yd

D. 3,850 yd

42. Simplify the expression below.

$$(-7x + 4) - (2x - 8)$$

F. $-5x - 4$

G. $-5x + 12$

H. $-9x - 4$

I. $-9x + 12$

Course 2 Benchmark Test – End of Year (continued)

43. The table shows the number of different types of rides at an amusement park. Which type of data display would be best show the number of items in specific categories?

Type of Ride	Number
Water Slides	9
Rollercoasters	14
Spinning Rides	5
Funhouses	4

A. bar graph

B. circle graph

C. line graph

D. line plot

44. SHORT ANSWER What is the surface area of the rectangular prism shown below?

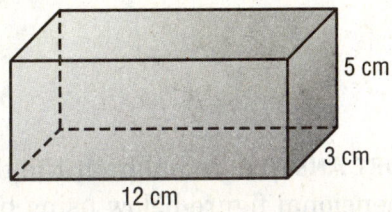

45. Angles C and E are supplementary. If $m\angle C = 77°$, what is the measure of angle E?

F. 13°

G. 77°

H. 103°

I. 113°

46. How much simple interest would be earned on an investment of $16,000 if the money is invested for 20 years at an annual interest rate of 5.25%?

A. $840

B. $16,800

C. $16,840

D. $32,800

47. A muffin recipe calls for 8 cups of flour and yields 24 muffins. If Natalie wants to make 60 muffins, how much flour will she need?

F. 180 cups

G. 24 cups

H. 20 cups

I. 3.2 cups

48. Which number line shows the solution to the inequality below?

$$-4g < 4$$

A. (open circle at 3, shaded left)

B. (open circle at -1, shaded right)

C. (open circle at 3, shaded right)

D. (open circle at -1, shaded left)

Course 2 Benchmark Test – End of Year (continued)

49. What is the area of the figure below? Use 3.14 for π. Round to the nearest tenth.

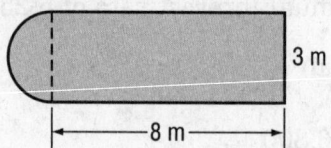

F. 24.0 m²

G. 27.5 m²

H. 31.1 m²

I. 38.1 m²

50. Christy drove 132 miles in $2\frac{3}{4}$ hours. What was her average speed in miles per hour?

A. 48 miles per hour

B. 46 miles per hour

C. 44 miles per hour

D. 42 miles per hour

51. The square pyramid has base side lengths of 12 centimeters and a slant height of 15 centimeters. What is the total surface area of the pyramid?

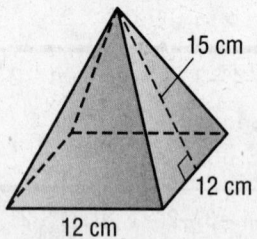

F. 720 in²

G. 640 in²

H. 504 in²

I. 360 in²

52. Suppose the length of each side of a square is decreased by 4 feet. If the perimeter of the square is now 32 feet, what was the original length of each side?

A. 48 ft

B. 44 ft

C. 16 ft

D. 12 ft

53. Which of the following is the simplest form of the algebraic expression shown below?

$$-11g + 5 + 6g - 2$$

F. $-6g + 4$

G. $-6g + 5$

H. $-5g + 5$

I. $-5g + 3$

54. SHORT ANSWER Jamal built the three-dimensional figure below using blocks.

front

Sketch the front, side, and top views of the figure.

Course 2 Benchmark Test – End of Year (continued)

55. Which of the following represents two dependent events?

A. drawing a card from a deck, not replacing it, and drawing another card

B. rolling a number cube and flipping a coin

C. drawing a card from a deck, replacing it, and drawing another card

D. rolling two numbers cubes

56. Which of the following is a possible cross section of a rectangular prism?

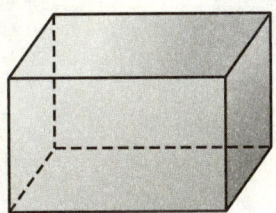

F. rectangle

G. oval

H. triangle

I. trapezoid

57. What is the solution to the equation?

$$4(x + 1) = -16$$

A. −3

B. −5

C. −63

D. −65

58. SHORT ANSWER Find the volume and surface area of the composite figure shown below if the figure is built with unit cubes.

59. Which operation should be performed last to solve the inequality below?

$$-7x + 4 > -10$$

F. add 4 to each side

G. subtract 4 from each side

H. multiply each side by −7 and reverse the inequality symbol

I. divide each side by −7 and reverse the inequality symbol

60. What is the product of the expression?

$$-7(-3)$$

A. −21

B. −10

C. 10

D. 21

Chapter 1 Answer Key

Are You Ready?—Review
Page 1

1. $\dfrac{4}{7}$
2. $\dfrac{1}{2}$
3. $\dfrac{1}{6}$
4. $\dfrac{7}{6}$
5. $\dfrac{2}{7}$
6. $\dfrac{1}{4}$
7. $\dfrac{3}{2}$
8. $\dfrac{7}{3}$

Are You Ready?—Practice
Page 2

1. $\dfrac{7}{1}$
2. $\dfrac{35}{11}$
3. $\dfrac{5}{1}$
4. $\dfrac{1}{7}$
5. $\dfrac{5}{7}$
6. $\dfrac{11}{35}$
7. $\dfrac{11}{7}$
8. $\dfrac{1}{5}$
9. $\dfrac{7}{5}$
10. $\dfrac{7}{11}$
11. yes; $\dfrac{11}{14} = \dfrac{22}{28}$
12. no; $\dfrac{3}{10} \neq \dfrac{4}{11}$
13. yes; $\dfrac{6}{8} = \dfrac{18}{24}$

Course 2 • Chapter 1 Ratios and Proportional Reasoning

Chapter 1 Answer Key

Are You Ready?—Apply
Page 3

1. **TRAVEL** Kimberly traveled to her friend's house and went 300 miles in 5 hours. On her way home she took a different route and traveled 420 miles in 7 hours. Are these ratios equivalent?

    yes; $\dfrac{300}{5} = \dfrac{420}{7}$

2. **TOMATOES** On Monday Janine picked 20 tomatoes off 4 tomato plants. On Thursday she picked 15 tomatoes off 3 tomato plants. Determine whether the ratios are equivalent.

    yes; $\dfrac{20}{4} = \dfrac{15}{3}$

3. **BASKETBALL** Daniel's basketball team won 23 games and lost 8 games. Write the ratio of wins to losses in simplest form.

    $\dfrac{23}{8}$

4. **MUSIC** Mr. Jansen listened to 8 songs in 28 minutes. He later listened to 5 songs in 21 minutes. Determine whether the ratios are equivalent.

    no; $\dfrac{8}{28} \neq \dfrac{5}{21}$

5. **MOVIE ATTENDANCE** Friday night's movie attendance is shown in the table. Write the ratio of males to females in simplest form.

Friday's Movie Attendance	
Males	58
Females	72

    $\dfrac{29}{36}$

6. **MOVIE ATTENDANCE** Use the table in Exercise 5 to write the ratio of females to the total number of people attending Friday night's movie in simplest form.

    $\dfrac{36}{65}$

Chapter 1 Answer Key

Diagnostic Test
Page 4

1. $\dfrac{25}{26}$
2. $\dfrac{25}{24}$
3. $\dfrac{25}{23}$
4. $\dfrac{26}{25}$
5. $\dfrac{26}{23}$
6. $\dfrac{24}{25}$
7. $\dfrac{24}{23}$
8. $\dfrac{23}{25}$
9. $\dfrac{23}{26}$
10. $\dfrac{23}{24}$
11. yes; $\dfrac{12}{36} = \dfrac{4}{12}$
12. no; $\dfrac{6}{10} \neq \dfrac{7}{11}$
13. yes; $\dfrac{4}{8} = \dfrac{7}{14}$

Pretest
Page 5

1. 25 mph
2. 3 laps per minute
3. $3.50 per pound
4. $x = 10$ m
5. 2
6. 78
7. 10
8. 7
9. yes
10. no

Course 2 • Chapter 1 Ratios and Proportional Reasoning

Chapter 1 Answer Key

Chapter Quiz
Page 6

1. 11.61 m/s
2. 8 pounds per dog
3. 16 oz
4. $\frac{3}{8}$
5. 9
6. 20
7. 85.1 ft/s
8. yes; Sample answer: The earnings to time ratios are all equal.
9. no; Sample answer: the cost to the number of bottles ratios are not equal.
10. 27 glasses

Vocabulary Test
Page 7

1. proportional
2. equivalent
3. equivalent ratios
4. cross products
5. constant rate of change
6. rate of change
7. direct variation
8. Sample answer: the ratio of vertical change to horizontal change
9. Sample answer: The rate for one unit of a given quantity

Chapter 1 Answer Key

Student Recording Sheet, Page 10
Use this recording sheet with the Standardized Test Practice.

Fill in the correct answer. For gridded-response questions, write your answers in the boxes on the answer grid and fill in the bubbles to match your answers.

1. Ⓐ Ⓑ ● Ⓓ

2. ● Ⓖ Ⓗ Ⓘ

3. 37.08

4. the number of students in her class

5. Ⓐ ● Ⓒ Ⓓ

6. ● Ⓖ Ⓗ Ⓘ

7. 286.44 mi

8. Ⓐ Ⓑ ● Ⓓ

9. ● Ⓖ Ⓗ Ⓘ

10. 3°, The temperature increased 3°F per hour.

11. Ⓐ Ⓑ ● Ⓓ

12. Ⓕ ● Ⓗ Ⓘ

13. 190 people

Extended Response
Record your answers for Exercise 14 on the back of this paper.

Part A   6 in./min

Part B   Sample answer: The point (0, 0) represents a water level of 0 at 0 minutes. The point (1, 6) represents 6 inches of water in the bathtub at 1 minute.

Chapter 1 Answer Key
Extended-Response Test, Page 11
Sample Answers

In addition to the scoring rubric, the following sample answers may be used as guidance in evaluating extended response assessment items.

1. Sample answers: 4:9 and 12:27; 3:8 and 9:24; 3:5 and 12:20

2. A proportion is an equation that shows that two ratios are equivalent.

3a. $100

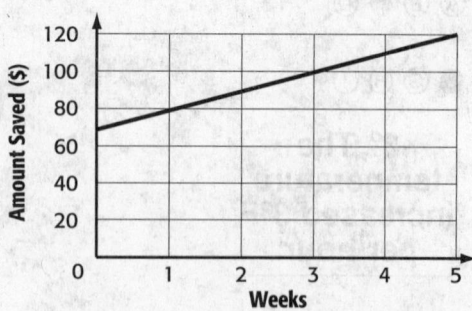

3b. the amount saved each week, $10

3c. No, the ratios of the amount saved to the week numbers are not equal.

Chapter 1 Answer Key

Test, Form 1A
Page 13

1. __B__

2. __I__

3. __B__

4. __G__

5. __C__

6. __H__

7. __B__

Test, Form 1A *(continued)*
Page 14

8. __H__

9. __B__

10. __F__

Chapter 1 Answer Key

Test, Form 1B
Page 15

1. C
2. I
3. B
4. G
5. C
6. G
7. A

Test, Form 1B *(continued)*
Page 16

8. F
9. C
10. I

Chapter 1 Answer Key

Test, Form 2A
Page 17

1. B
2. F
3. C
4. F
5. A
6. I

Test, Form 2A *(continued)*
Page 18

7. C
8. G
9. $9.00
10. 138 miles
11. 406 miles
12. 740 words

Chapter 1 Answer Key

Test, Form 2B
Page 19

1. D

2. F

3. C

4. G

5. B

6. I

Test, Form 2B *(continued)*
Page 20

7. A

8. G

9. $7.80

10. 440 miles

11. 285 miles

12. 35 purses

Chapter 1 Answer Key

Test, Form 3A
Page 21

1. __40 oz__

2. __the rate of which money was saved__

3. __25 km__

4. __No; $\frac{6 \text{ bottles}}{\$4} = \frac{1.5 \text{ bottles}}{\$1}$ and $\frac{18 \text{ bottles}}{\$10} = \frac{1.8 \text{ bottles}}{\$1}$; these rates are not equivalent.__

5. __15.6__

6. __72.8__

7. __25.35__

8. __58.67 ft/s__

Test, Form 3A (continued)
Page 22

9. __$2\frac{1}{2}°$ per hour__

10. __Yes, the graph shows a straight line that passes through the origin.__

11. __7 days__

12. __$3.40__

13. __405__

14. __Jaime; since the slope of the line is 7 and the table shows a rate of $8 per hour, Jaime makes more.__

Course 2 • Chapter 1 Ratios and Proportional Reasoning

Chapter 1 Answer Key

Test, Form 3B
Page 23

1. 10 oz

2. the rate at which money was saved

3. 6 km

    No; $\frac{6 \text{ bottles}}{\$4.80} = \frac{1.25 \text{ bottles}}{\$1}$ and $\frac{18 \text{ bottles}}{\$22.50} = \frac{0.8 \text{ bottles}}{\$1}$; these rates are not equivalent.

4. _____

5. 2.25

6. 6.88

7. 18.75

8. 299.33 ft/s

Test, Form 3B *(continued)*
Page 24

9. 2 degrees per hour.

10. Yes, the graph shows a straight line that passes through The origin.

11. 4 days

12. $3.96

13. 324

14. Crystal; since the slope of the line is 12 and the table shows a rate of $10 per hour, Crystal makes more.

Chapter 2 Answer Key

Are You Ready?—Review
Page 25

1. 0.5
2. 0.06
3. 1.17
4. 0.13
5. 0.0125
6. 0.325
7. 2.3
8. 0.74
9. 0.279
10. 0.085
11. 6.12
12. 0.039

Are You Ready?—Practice
Page 26

1. 16
2. 5.4
3. 6.3
4. 105
5. 0.4
6. 0.08
7. 1.75
8. 0.805
9. 0.092
10. 4.32
11. 0.55
12. 14%
13. 2%
14. 27.5%
15. 7.6%
16. 230%
17. 585%
18. 3.7%

Course 2 • Chapter 2 Percents

Chapter 2 Answer Key

Are You Ready?—Apply
Page 27

1. **TRACK** A track team won first place in 0.91 of the meets they competed in this year. What percent of the meets did the team win first place?   **91%**

2. **FAVORITE COLOR** Mr. McGuirk surveyed the students in his math class and found that 45% of them said their favorite color was red. What decimal represents this amount?
   0.45

3. **TAX** A state sales tax is 6.75%. Write this percent as a decimal.   **0.0675**

4. **GOLF** When Akil golfs, 85% of the time he hits the green on his second shot. What decimal represents this amount?   **0.85**

5. **SALES TAX** The sales tax on Caden's groceries was 6.5%. Write this percent as a decimal.   **0.065**

6. **BANK ACCOUNT** The interest rate Calvin is earning on his bank account is 4%. Write this percent as a decimal.
   0.04

Chapter 2 Answer Key

Diagnostic Test
Page 28

1. 15
2. 87.5
3. 7.44
4. 11.2
5. 15 questions
6. 0.3
7. 0.06
8. 2.15
9. 0.632
10. 0.037
11. 0.865
12. 7.2
13. 23%
14. 9%
15. 31.8%
16. 428%
17. 1.8%
18. 94%
19. 368%
20. 17.1%

Pretest
Page 29

1. 7.2
2. 20
3. 94.5
4. 0.7
5. 72%
6. 20%
7. 60
8. 48
9. $800
10. 17% increase
11. 20% decrease
12. 67% increase
13. $9,450
14. $149.80
15. $238.50

Course 2 • Chapter 2 Percents

A15

Chapter 2 Answer Key

Chapter Quiz
Page 30

1. 14
2. 25.9
3. 40.8
4. 24
5. Sample answer: $\frac{1}{10} \cdot 40 = 4$
6. Sample answer: $\frac{1}{4} \cdot 60 = 15$
7. $n = 0.08 \cdot 50$; 4
8. $52 = n \cdot 260$; 20%
9. $30 = 0.75 \cdot n$; 40
10. $n = 0.15 \cdot 24$; 3.6
11. 45 questions
12. 10%

Vocabulary Test
Page 31

1. f
2. b
3. c
4. g
5. d
6. h
7. e
8. e
9. a
10. i
11. Sample answer: percent proportion compares part of a quantity to the whole quantity using a percent.
12. Sample answer: gratuity is the amount you pay extra to show you appreciated the service.

Chapter 2 Answer Key

Student Recording Sheet, Page 34

Use this recording sheet with the Standardized Test Practice pages.

Fill in the correct answer. For gridded-response questions, write your answers in the boxes on the answer grid and fill in the bubbles to match your answers.

1. Ⓐ Ⓑ ● Ⓓ
2. Ⓕ ● Ⓗ Ⓘ
3. ● Ⓑ Ⓒ Ⓓ
4. 15.30
5. 95
6. Ⓕ ● Ⓗ Ⓘ
7. Ⓐ ● Ⓒ Ⓓ
8. Ⓕ Ⓖ ● Ⓘ
9. 68
10. 4%
11. Ⓐ Ⓑ Ⓒ ●
12. Ⓕ Ⓖ Ⓗ ●
13. Ⓐ Ⓑ ● Ⓓ

Extended Response

Record your answers for Exercise 14 on the back of this paper.

Part A 4%

Part B Sample answer: you can find 10% of $110 and subtract the amount from $110; $110 − $11 = $99. You can also subtract 10% from 100% and multiply the new percent by $110; 90% × $110 = $99.

Part C Yes, Company B is $2.92 cheaper per month for new customers.

Course 2 • Chapter 2 Percents

Chapter 2 Answer Key

Extended-Response Test, Page 35
Sample Answers

In addition to the scoring rubric, the following sample answers may be used as guidance in evaluating extended response assessment items.

1. $0.20 \cdot 2000 = n$
   $400 = n$

2. **a.** $65n = 65 - 41.99$ Write an equation.
   $65n = 23.01$ Subtract.
   $n = \dfrac{23.01}{65}$ Divide both sides by 65.
   $n = 0.354$ Simplify.
   $n = 35.4\%$ Change the decimal to a percent.

   b. $16.95 \cdot 0.22 = n$ Write an equation to find the amount of discount.
   $3.73 \approx n$ Multiply.
   $16.95 - 3.73 = n$ Find the sale price.
   $n = 13.22$ Subtract.
   $13.22 \cdot 0.065 = t$ Write an equation to find the amount of tax.
   $0.86 \approx t$ Multiply.
   $13.22 + 0.86 = t$ Find the sale price with tax.
   $\$14.08 = t$ Add.

3. **a.** $I = Prt$ Write the interest formula.
   $I = 1{,}200 \cdot 0.08 \cdot \dfrac{1}{2}$ Fill in the principal, rate, and time given in the formula.
   $I = \$48$ Multiply.

   b. First year
   $I = Prt$ Write the interest formula.
   $I = 900 \cdot 0.06 \cdot 1$ Fill in the principal, rate, and time given in the formula.
   $I = 54$ Multiply.
   $900 + 54 = 954$ At the end of the first year, the new balance is $954.

   Second year
   $I = Prt$ Write the interest formula.
   $I = 954 \cdot 0.06 \cdot 1$ Fill in the principal, rate, and time given in the formula.
   $I = 57.24$ Multiply.
   $954 + 57.24 = 1{,}011.24$ At the end of the second year, the new balance is $1,011.24.

   Third year
   $I = Prt$ Write the interest formula.
   $I = 1{,}011.24 \cdot 0.06 \cdot 1$ Fill in the principal, rate, and time given in the formula.
   $I \approx 60.67$ Multiply.
   $1{,}011.24 + 60.67 = 1{,}071.91$ At the end of the third year, the new balance is $1,071.91.

   Fourth year
   $I = Prt$ Write the interest formula.
   $I = 1{,}071.91 \cdot 0.06 \cdot 1$ Fill in the principal, rate, and time given in the formula.
   $I \approx 64.31$ Multiply.
   $1{,}071.91 + 64.31 = 1{,}136.22$ At the end of the fourth year, the new balance is $1,136.22.

Chapter 2 Answer Key

Test, Form 1A
Page 37

1. B
2. H
3. A
4. H
5. C
6. G
7. D
8. I
9. A
10. I

Test, Form 1A *(continued)*
Page 38

11. B
12. H
13. C
14. H
15. C
16. I
17. C
18. F
19. D
20. G

Chapter 2 Answer Key

Test, Form 1B
Page 39

1. C
2. I
3. A
4. G
5. A
6. H
7. A
8. F
9. C
10. H

Test, Form 1B *(continued)*
Page 40

11. C
12. H
13. D
14. F
15. C
16. G
17. A
18. F
19. B
20. H

Chapter 2 Answer Key

Test, Form 2A
Page 41

1. B
2. I
3. A
4. I
5. B
6. H
7. D
8. H
9. C

Test, Form 2A *(continued)*
Page 42

10. G
11. $1.50
12. 37% increase
13. 38% decrease
14. 17% increase
15. $25.50
16. $45.60
17. $63.89
18. $157.50
19. $18.40

Chapter 2 Answer Key

Test, Form 2B
Page 43

1. C
2. F
3. A
4. H
5. C
6. H
7. D
8. G
9. C

Test, Form 2B *(continued)*
Page 44

10. F
11. $1.34
12. 21% increase
13. 36% decrease
14. 24% increase
15. $72.50
16. $58.50
17. $68.52
18. $71.20
19. $24.15

Chapter 2 Answer Key

Test, Form 3A
Page 45

1. 24
2. 100
3. 12.5%
4. $0.6 \times 50 = 30$
5. $\frac{2}{5} \times 130 = 52$
6. $2 \times 20 = 40$
7. $15 = n \cdot 48$; 31.3%
8. $42 = 0.35 \cdot w$; $w = 120$
9. 39 questions
10. 29%
11. 41%
12. Emilio; $1.25 \cdot 0.08 = 0.10 + 1.25 = 1.35$; $1.35 \times 30 = 40.5$; $1.40 \cdot 0.05 = 0.07 + 1.4 = 1.47$; $1.47 \times 26 = 38.22$; Emilio made $40.50; Regina made $38.22.

Test, Form 3A *(continued)*
Page 46

13. 67%; decrease
14. 9%; increase
15. 37%; increase
16. 14%; increase
17. 13%
18. $8.34
19. $25.50
20. $2,662.50
21. $14.89
22. $53.20
23. $10.47
24. $11.48
25. $3,549.58

Chapter 2 Answer Key

Test, Form 3B
Page 47

1. 15.2
2. 9,400
3. 6.25%
4. 40
5. 22
6. $3 \cdot 25 = 75$
7. $14 = n \cdot 70$; 20%
8. $0.65 \cdot w = 39$; $w = 60$
9. 30 books
10. 29%
11. 37%
12. Suzanne; Suzanne's: $2.00 \times 0.05 + 2.00 = 2.10 \times 34 = \$71.40$; Michael's: $2.25 \times 0.04 + 2.25 = 2.34 \times 30 = \$70.20$

Test, Form 3B *(continued)*
Page 48

13. 83% decrease
14. 44% increase
15. 150% increase
16. 121% increase
17. 28%
18. $17.19
19. $30.09
20. $13.59
21. $26.38
22. $393.00
23. $5.70
24. $6.19
25. $2,561.50

Chapter 3 Answer Key

Are You Ready?—Review
Page 49

1. 4
2. 51
3. 41
4. 8
5. 25
6. 2
7. 9
8. 15
9. 24
10. 14

Are You Ready?—Practice
Page 50

1. 3
2. 60
3. 10
4. 42
5. 2
6. 6
7. 10
8. 20
9. (2, 6)
10. (3, 1)
11. (4, 4)
12. (7, 2)
13. (6, 8)
14. (1, 2)
15. (5, 3)

Course 2 • Chapter 3 Integers

Chapter 3 Answer Key

Are You Ready?—Apply
Page 51

1. **BASEBALL** John had 30 baseball cards. He gave 14 cards to Mike and 7 to Jeff. How many baseball cards does John have left? **9**

2. **PARTIES** Louise is having a party. She bought 96 pieces of chicken. If she planned to have 3 pieces for each guest, how many people could she invite? **32**

3. **FUNDRAISER** The soccer team collected soda cans for a fundraiser. They had 175 cans and found 58 more. The next day, they turned in 97 cans. How many cans do they have left? **136**

4. **CORN** Mr. Rodriguez planted 22 rows of corn. There were 15 plants in each row. How many corn plants did he put in his garden? **330**

5. **MAPS** The graph below shows the locations of four places. What are the coordinates of the skating rink? **(4, 3)**

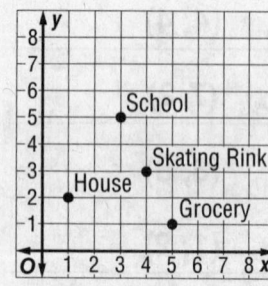

6. **MAPS** Refer to Exercise 5. The library is located at (2, 4). Which location on the map is closest to the library? **school**

A26   Course 2 • Chapter 3 Integers

Chapter 3 Answer Key

Diagnostic Test
Page 52

1. 43
2. 7
3. 9
4. 14
5. 11
6. 3
7. 19
8. 33

9. (7, 4)
10. (2, 6)
11. (3, 1)
12. (6, 2)
13. (9, 10)
14. (1, 9)
15. (8, 7)

Pretest
Page 53

1.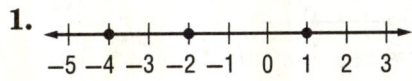
2.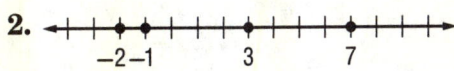
3. 11
4. 10
5. 3
6. −8
7. 4
8. −5
9. −18
10. −2
11. 27
12. 7
13. −5
14. −18
15. −$12

Course 2 • Chapter 3 Integers

A27

Chapter 3 Answer Key

Chapter Quiz
Page 54

1.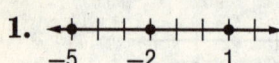

2. (number line with points at −3, 2, 4)

3. −15
4. 24
5. 3
6. 9
7. 15
8. 4
9. 5
10. −1
11. −10
12. 11
13. −38 feet

Vocabulary Test
Page 55

1. absolute value
2. integer
3. additive inverse
4. negative integer
5. opposites
6. graph
7. positive integer

Chapter 3 Answer Key

Student Recording Sheet, Page 58

Use this recording sheet with the Standardized Test Practice pages.

Fill in the correct answer. For gridded-response questions, write your answers in the boxes on the answer grid and fill in the bubbles to match your answers.

1. Ⓐ Ⓑ ● Ⓓ

2. Ⓕ ● Ⓗ Ⓘ

3. 3780

4. −25

5. Ⓐ Ⓑ ● Ⓓ

6. ● Ⓖ Ⓗ Ⓘ

7. Ⓐ ● Ⓒ Ⓓ

8. 65

9. $(-12,000) + 4,411 = (-7,589)$; So, Larry owes his grandfather $7,589.

10. ● Ⓖ Ⓗ Ⓘ

11. Ⓐ Ⓑ ● Ⓓ

12. Sample answer: (−3, 1)

Extended Response

Record your answers for Exercise 13 on the back of this paper.

Part A   2003; −26,000 from 2002

Part B   10,000 −14,000; −4,000

Chapter 3 Answer Key
Extended-Response Test, Page 59
Sample Answers

In addition to the scoring rubric, the following sample answers may be used as guidance in evaluating extended response assessment items.

1. **a.** The absolute value of a number is the distance a number is from zero on the number line; $|-3| = 3$.

   b. $-3 + (+2)$

   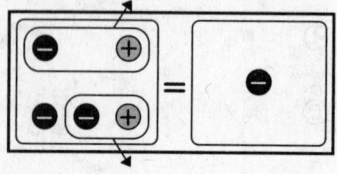

   Glen's score after two rounds is -1.

   c. $-1 + (-3) = -4$

2. **a.** $\dfrac{C - S}{5}$

   b. $\dfrac{C - S}{5} = \dfrac{13{,}000 - 43{,}500}{5} = \dfrac{-30{,}500}{5} = -6{,}100$; $6,100

3. Sample answer: To subtract an integer from another, you can add its opposite. $8 - 3$ is the same as $8 + (-3)$.

Chapter 3 Answer Key

Test, Form 1A
Page 61

1. A
2. H
3. B
4. I
5. C
6. F
7. C
8. G
9. C
10. H

Test, Form 1A *(continued)*
Page 62

11. C
12. G
13. A
14. H
15. B
16. I
17. B
18. H
19. D
20. H

Chapter 3 Answer Key

Test, Form 1B
Page 63

1. D
2. I
3. A
4. I
5. C
6. G
7. D
8. F
9. A
10. I

Test, Form 1B *(continued)*
Page 64

11. B
12. H
13. A
14. H
15. B
16. H
17. B
18. I
19. C
20. F

Chapter 3 Answer Key

Test, Form 2A
Page 65

1. D
2. F
3. D
4. H
5. B
6. F
7. C
8. G
9. C

Test, Form 2A *(continued)*
Page 66

10. H
11. C
12. I
13. A
14. F
15. D
16. 13
17. 12
18. −1
19. 8
20. −32
21. 1
22. 56
23. −12
24. (number line with points at −4, −2, and 1 on scale −4 to 2)

Chapter 3 Answer Key

Test, Form 2B
Page 67

1. B
2. H
3. A
4. I
5. C
6. H
7. B
8. F
9. C

Test, Form 2B *(continued)*
Page 68

10. G
11. C
12. F
13. B
14. F
15. D
16. 11
17. −20
18. −1
19. 17
20. −3
21. 1
22. −15
23. −2
24. number line with points at −1, 1, and 4

Chapter 3 Answer Key

Test, Form 3A
Page 69

1. 45
2. −5
3. −15 yards
4. 11
5. 10
6. 5
7. (number line with points at −5, −3, and 2 on a scale from −6 to 3)
8. −2°F per minute
9. 7,218°C
10. 22
11. −28
12. −13
13. −11
14. −8
15. −37
16. −88
17. −5

Test, Form 3A *(continued)*
Page 70

18. 6
19. 9
20. 2
21. −35
22. 24
23. $300; (−$25)(8) or −$200 is how much Trent spent in 8 weeks. $500 − $200 = $300.
24. 21
25. −5
26. 4
27. 18
28. −2
29. 2
30. −24
31. −1

Course 2 • Chapter 3 Integers

A35

Chapter 3 Answer Key

Test, Form 3B
Page 71

1. 300
2. −13
3. $45
4. 13
5. 8
6. 2
7. [number line with points at −6, −3, and 2, marked between −6 and 3]
8. −3°F per minute
9. $340
10. 29
11. −27
12. −17
13. −9
14. 4
15. −20
16. 9
17. 25

Test, Form 3B *(continued)*
Page 72

18. −12
19. −17
20. 2
21. −42
22. 15
23. $480; (−$30)(9) or −$270 is how much Jean spent in 9 weeks. $750 − $270 = $480.
24. 13
25. −8
26. 7
27. 30
28. −4
29. −3
30. −40
31. −6

Chapter 4 Answer Key

Are You Ready?—Review
Page 73

Are You Ready?—Practice
Page 74

1. 0.7
2. 0.49
3. 0.29

4.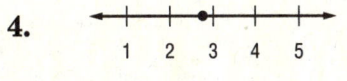

5.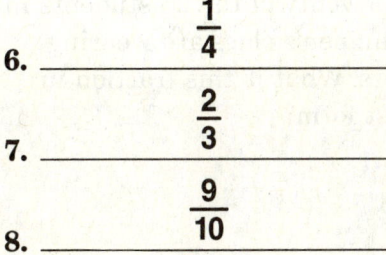

1. $\dfrac{3}{4}$
2. $\dfrac{4}{5}$
3. $\dfrac{2}{3}$
4. $\dfrac{1}{2}$
5. $\dfrac{7}{10}$
6. 1
7. $1\dfrac{1}{3}$
8. simplified

6. $\dfrac{1}{4}$
7. $\dfrac{2}{3}$
8. $\dfrac{9}{10}$

9. Economy Nuts

10.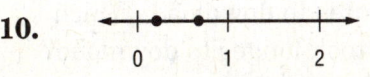

Course 2 • Chapter 4 Rational Numbers

A37

Chapter 4 Answer Key

Are You Ready?—Apply
Page 75

1. SHOPPING Marcia spent $0.76 for a cookie. Ken spent $0.67 for a candy bar. Who spent more? **Marcia**	**2. FISHING** Bruce's fish weighed 0.96 pound while Mosey's fish weighed 1.16 pounds. Whose fish weighed more? **Mosey's fish**
3. SHOES Twenty of the 25 students in Mrs. Thigpen's class are wearing sneakers. What is this fraction in simplest form? $\frac{4}{5}$	**4. FOOTBALL** Hill ran for 80 of his team's 120 rushing yards in Saturday's football game. Sheets ran for 60 of his team's 80 rushing yards. What are these fractions in simplest form? Hill $\frac{2}{3}$, Sheets $\frac{3}{4}$
5. SOFTWARE Leslie downloaded two software programs onto her computer. The first program took 5.76 minutes to download while the second took 5.06 minutes to download. Which program took longer to download? **the first program**	**6. HIKING** Morgan stopped to rest after hiking 1.8 kilometers. Her mother rested after 2.1 kilometers. Who hiked farther? **Morgan's mother**

Chapter 4 Answer Key

Diagnostic Test
Page 76

1. 2.63
2. 0.51

3.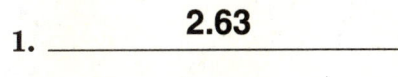
4.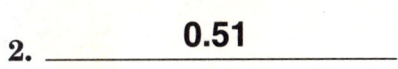

5. $\dfrac{6}{7}$
6. $\dfrac{2}{5}$
7. simplified
8. $\dfrac{1}{4}$

9. $\dfrac{1}{3}$

10.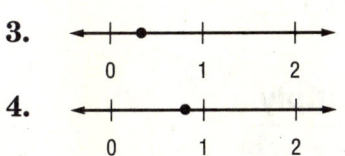

Pretest
Page 77

1. 0.625
2. $0.\overline{18}$

3. $\dfrac{3}{5}$
4. $1\dfrac{1}{10}$
5. $3\dfrac{1}{2}$
6. $1\dfrac{7}{10}$
7. $1\dfrac{1}{2}$
8. $2\dfrac{1}{2}$

9. >

10. 25.4

11. 0.64

12. 6,624.45

13. $\dfrac{19}{40}$

Course 2 • Chapter 4 Rational Numbers

Chapter 4 Answer Key

Chapter Quiz
Page 78

1. 2.625
2. $0.\overline{7}$
3. 0.75
4. <
5. =
6. $1\frac{1}{7}$
7. $1\frac{1}{3}$
8. $-\frac{4}{5}$
9. $\frac{7}{16}$
10. $\frac{5}{12}$ hour

Vocabulary Test
Page 79

1. true
2. false; multiply
3. true
4. true
5. false; terminating decimal
6. true
7. false; add
8. false; divide
9. Sample answer: The least common denominator of two fractions is the smallest number that is a common multiple of their denominators.
10. Sample answer: A rational number is any number that can be written as a fraction.

Chapter 4 Answer Key

Student Recording Sheet, Page 82

Use this recording sheet with the Standardized Test Practice pages.

Fill in the correct answer. For gridded-response questions, write your answers in the boxes on the answer grid and fill in the bubbles to match your answers.

1. $n = 2\frac{1}{4} - \frac{3}{4}$

2. D

3. H

4. 30

5. 270

6. C

7. H

8. D

9. 18

10. H

11. 30

12. 7/8

13. 7 cups of flour

14. Part A $l = 35 \div 1\frac{1}{4}$
    Part B 28 loads
    Part C 84 loads

Extended Response

Record your answers for Exercise 14 on the back of this paper.

Course 2 • Chapter 4 Rational Numbers

Chapter 4 Answer Key
Extended-Response Test, Page 83
Sample Answers

In addition to the scoring rubric, the following sample answers may be used as guidance in evaluating extended response assessment items.

1. **a.** Method 1: $2\frac{1}{4} = 1\frac{5}{4}$

   $\underline{-1\frac{1}{2} = 1\frac{2}{4}}$

   $\frac{3}{4}$ lb

   Method 2:

   $2\frac{1}{4} =$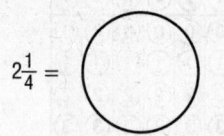

   $-1\frac{1}{2} =$

    lb

   b. number of groups × size of group

   $2\frac{1}{2}$ × $\frac{1}{2}$

   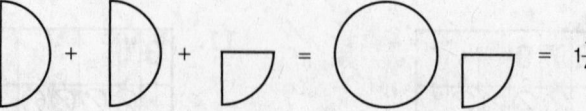 $= 1\frac{1}{4}$

   c. $2\frac{1}{2} \times \frac{1}{2} = \frac{5}{2} \times \frac{1}{2} = \frac{5}{4} = 1\frac{1}{4}$ gal

2. **a.** Sample answer: Sally bought $2\frac{1}{2}$ lb of ground chuck for $6.25. How much did she pay per pound for the ground chuck?

   b. $\$6.25 \div 2\frac{1}{2} = \$6.25 \div \frac{5}{2}$

   $\phantom{\$6.25 \div 2\frac{1}{2}} = \$6.25 \times \frac{2}{5}$

   $\phantom{\$6.25 \div 2\frac{1}{2}} = \$2.50/\text{lb}$

3. Brand 1: $\frac{\$2.00}{1 \text{ lb } 9 \text{ oz}} = \frac{200¢}{25 \text{ oz}} = \frac{8¢}{\text{oz}}$

   Brand 2: $\frac{\$2.10}{1 \text{ lb } 14 \text{ oz}} = \frac{210¢}{30 \text{ oz}} = \frac{7¢}{\text{oz}}$

   Brand 2 is the better buy, since it has a lower price per ounce.

Chapter 4 Answer Key

Test, Form 1A
Page 85

1. C
2. H
3. A
4. H
5. B
6. G
7. C
8. H
9. B
10. I

Test, Form 1A *(continued)*
Page 86

11. C
12. H
13. A
14. H
15. B
16. I
17. B
18. I
19. C

Course 2 • Chapter 4 Rational Numbers

Chapter 4 Answer Key

Test, Form 1B
Page 87

1. B

2. G

3. B

4. H

5. A

6. G

7. B

8. I

9. C

10. I

Test, Form 1B *(continued)*
Page 88

11. A

12. H

13. A

14. G

15. D

16. F

17. D

18. F

19. C

Chapter 4 Answer Key

Test, Form 2A
Page 89

1. C
2. F
3. D
4. F
5. C
6. G
7. B
8. I
9. A
10. I

Test, Form 2A *(continued)*
Page 90

11. C
12. F
13. B
14. $2\frac{9}{10}$ miles
15. $5\frac{1}{4}$ cups
16. 13 in.
17. $8\frac{5}{8}$ ft²
18. $\frac{5}{12}$
19. 10.56
20. 1.60
21. 1,088.31
22. 0.16
23. 125.39
24. 26.67

Chapter 4 Answer Key

Test, Form 2B
Page 91

1. C
2. H
3. C
4. F
5. D
6. G
7. A
8. G
9. D
10. H

Test, Form 2B *(continued)*
Page 92

11. B
12. G
13. A
14. $1\frac{3}{10}$ miles
15. $7\frac{1}{2}$ cups
16. 10 in.
17. $19\frac{1}{6}$ sq. in.
18. $\frac{7}{12}$
19. 3.31
20. 26.71
21. 0.54
22. 2,721.6
23. 0.42
24. 2.11

Chapter 4 Answer Key

Test, Form 3A
Page 93

1. $0.\overline{8}$
2. -6.75
3. $-\dfrac{1}{50}$
4. $68\dfrac{1}{4}$
5. $>$
6. $<$
7. $\dfrac{1}{4}, \dfrac{10}{25}, 0.5$
8. $-\dfrac{2}{3}$
9. $-\dfrac{3}{28}$
10. $-1\dfrac{1}{20}$
11. $1\dfrac{7}{15}$
12. $9\dfrac{11}{12}$
13. $-\dfrac{2}{9}$
14. $2\dfrac{4}{5}$
15. $4\dfrac{1}{4}$

Test, Form 3A *(continued)*
Page 94

16. 2 miles per hour
17. $8\dfrac{1}{4}$ cups
18. $2\dfrac{1}{6}$ hours
19. $13\dfrac{1}{3}$ ft
20. 18 pieces
21. 5:30 P.M.; Sample answer: 245 ÷ 35 = 7; The family traveled 245 miles in 7 sets of half hours, or $3\dfrac{1}{2}$ hours. Adding $3\dfrac{1}{2}$ hours to 2:00 P.M. will make it 5:30 P.M.
22. 6.3
23. 13.62
24. 16.31
25. 4,445.28

Course 2 • Chapter 4 Rational Numbers

Chapter 4 Answer Key

Test, Form 3B
Page 95

1. $0.\overline{27}$
2. -7.6
3. $-\dfrac{3}{50}$
4. $86\dfrac{3}{4}$
5. $<$
6. $<$
7. $0.8, \dfrac{7}{8}, \dfrac{15}{16}$
8. $-\dfrac{5}{7}$
9. $-\dfrac{5}{21}$
10. $-\dfrac{3}{4}$
11. $4\dfrac{3}{4}$
12. $4\dfrac{2}{5}$
13. $-\dfrac{18}{35}$
14. $2\dfrac{1}{2}$
15. $10\dfrac{1}{2}$

Test, Form 3B (continued)
Page 96

16. $\dfrac{21}{10}$ miles per hour
17. $6\dfrac{3}{4}$ tsp
18. $2\dfrac{1}{2}$ hours
19. $10\dfrac{3}{8}$ feet
20. 30 pieces
21. 7:30 P.M.; Sample answer: 225 ÷ 25 = 9; The family traveled 225 miles in 9 sets of half hours, or $4\dfrac{1}{2}$ hours. Adding $4\dfrac{1}{2}$ hours to 3:00 P.M. will make it 7:30 P.M.
22. 14
23. 13.43
24. 3.27
25. 0.26

Chapter 5 Answer Key

Are You Ready?—Review
Page 97

1. 8.6
2. 56.7
3. 19.5
4. 38.4
5. 7.14
6. 3.52
7. 6.67
8. 13.68
9. 26.13
10. 64.6

Are You Ready?—Practice
Page 98

1. 43.5
2. 52.1
3. 13.7
4. 24.91
5. 60.52
6. 52.19
7. $84.84
8. 1.2 mi
9. 17.2
10. 22.2
11. 13.8
12. 1.9
13. 37.72
14. 5.7
15. $1.75

Course 2 • Chapter 5 Expressions

Chapter 5 Answer Key

Are You Ready?—Apply
Page 99

1. **FOOD** Jasmine bought the items shown in the table. What was the total cost? **$4.35**

Item	Cost ($)
burrito	2.65
lemonade	1.70

2. **MONEY** Ethan had $15. He spent $13.74 on art supplies. How much money did Ethan have left? **$1.26**

3. **ENTERTAINMENT** Three friends spent $26.25 on movie tickets. How much does each ticket cost? **$8.75**

4. **PUPPY** The weight of a puppy is shown in the table. How much more did the puppy weigh in Week 8 than in Week 4? **7.6 lb**

Week	Weight (lb)
4	5.8
8	13.4

5. **JOBS** Ariana earns $12.35 per hour. What does she earn in one week if she works 18 hours? **$222.30**

6. **NUTRITION** A can of peas contains 3.5 servings and one serving is 0.5 cup. How many cups of peas are in the can? **1.75 cups**

Chapter 5 Answer Key

Diagnostic Test
Page 100

1. 22.1
2. 64.46
3. 48.3
4. 65.97
5. 13.19
6. $28.98
7. 3.25 mi
8. 7.4
9. 12
10. 6.4
11. 91.02
12. 3.5
13. 3.8 min
14. 4.3 oz
15. $107.40

Pretest
Page 101

1. 7
2. 3
3. 5 is added to each term; 24, 29, 34
4. 14 is added to each term; 56, 70, 84
5. $16.80
6. Multiplicative (0)
7. Associative (+)
8. $4x + 28$
9. $-5y - 50$
10. $8x + 6$
11. $2x + 2$
12. $2x - 5$
13. $3x$
14. $4a$
15. $5d$

Course 2 • Chapter 5 Expressions

Chapter 5 Answer Key

Chapter Quiz
Page 102

1. 4
2. 8
3. 4
4. 4(9) + 4(1); 40
5. −30 − 5x
6. Identity (×)
7. Associative (+)
8. Commutative (+)
9. 6th day
10. add 7; 30, 37, 44
11. add 3; 62, 65, 68
12. add 0.4; 1.8, 2.2, 2.6
13. add 9; 117, 126, 135

Vocabulary Test
Page 103

1. arithmetic sequence
2. coefficient
3. term
4. algebraic expression
5. equivalent expressions
6. constant
7. simplest form
8. Like terms

Chapter 5 Answer Key

Student Recording Sheet, Page 106
Use this recording sheet with the Standardized Test Practice pages.

Fill in the correct answer. For gridded-response questions, write your answers in the boxes on the answer grid and fill in the bubbles to match your answers.

1. A ● C D

2. F ● H I

3. 15

4. A B ● D

5. 56

6. F G ● I

7. A ● C D

8. 45.5

9. F G ● I

10. A B C ●

11. F G ● I

12. $50s + 55m + 53t$

13. A ● C D

14. $4(12 + 8); 80$

15. F ● H I

Extended Response
Record your answers for Exercise 16 on the back of this paper.

Part A

Part B Multiply the term number by 6 and subtract 2.

Part C $6x - 2$

Course 2 • Chapter 5 Expressions

Chapter 5 Answer Key
Extended-Response Test, Page 107
Sample Answers

In addition to the scoring rubric, the following sample answers may be used as guidance in evaluating extended response assessment items.

1. $7p + 6(p \div q) - 2q = 7(6) + 6(6 \div 3) - 2(3)$ Replace p with 6 and q with 3.
   $= 7(6) + 6(2) - 2(3)$ Divide first since $6 \div 3$ is in parentheses.
   $= 42 + 12 - 6$ Multiply.
   $= 54 - 6$ Add from left to right, $42 + 12$.
   $= 48$ Subtract $54 - 6$.

2. **a.** Let s represent the number of student tickets purchased and a represent the number of adult tickets purchased; $25s + 5a$

   b. $25 \cdot 4 + 5 \cdot 6$; $130

3. 752, 756, 760, 764, 768
       +4   +4   +4   +4

   Each term is found by adding 4 to the previous term. The next term can be found by adding 4 to the last term.
   $768 + 4 = 772$

4. Add or subtract similar terms.

Chapter 5 Answer Key

Test, Form 1A
Page 109

1. C
2. G
3. C
4. G
5. B
6. H
7. A

Test, Form 1A *(continued)*
Page 110

8. H
9. D
10. H
11. A
12. G
13. A
14. F

Chapter 5 Answer Key

Test, Form 1B
Page 111

1. D
2. G
3. D
4. G
5. B
6. I
7. B

Test, Form 1B *(continued)*
Page 112

8. H
9. D
10. F
11. B
12. I
13. B
14. F

Chapter 5 Answer Key

Test, Form 2A
Page 113

1. C

2. F

3. B

4. I

5. B

6. H

7. A

Test, Form 2A *(continued)*
Page 114

8. I

9. C

10. F

11. 18

12. Associative

13. 3 units by (5x − 3) units

14. (x + 7) dollars

Course 2 • Chapter 5 Expressions

A57

Chapter 5 Answer Key

Test, Form 2B
Page 115

1. D

2. F

3. B

4. I

5. B

6. I

7. C

Test, Form 2B *(continued)*
Page 116

8. H

9. D

10. F

11. 18

12. Commutative

13. 9 units by (3x + 2) units

14. (x + 4) dollars

Chapter 5 Answer Key

Test, Form 3A
Page 117

1. __Identity__

2. __−2n − 18__

3. __Commutative__

4. __16x − 36__

5. __25, 29, 33__

6. __7a__

7. __−8x − 6__

8. __13x − 2__

Test, Form 3A *(continued)*
Page 118

9. __$29.25; 3($10 − $0.25) = 3 · 10 − 3 · 0.25__

10. __4a + 2__

11. __−28__

12. __17__

13. __30__

14. __$42; Sample answer: $9.50 + $15.50 = $25 and $11.75 + $5.25 = $17, $25 + $17 = $42.__

15. __The GCF of 7 and 5 is 1. If the GCF of the two terms is 1, the expression cannot be factored.__

Course 2 • Chapter 5 Expressions

A59

Chapter 5 Answer Key

Test, Form 3B
Page 119

1. Identity

2. $6w + 30$

3. Distributive

4. $-2x - 31$

5. 6.6, 7.2, 7.8

6. $3a$

7. $-5x - 10$

8. $15x - 3$

Test, Form 3B *(continued)*
Page 120

9. $15.50; 5($3 + $0.10) = 5(3) + 5(0.10)$

10. $4b + 3$

11. -21

12. 10

13. 32

14. $130; 24 + 48 + 36 + 22 = (24 + 36) + (48 + 22) = 60 + 70 = 130.$

15. The GCF of 13 and 10 is 1. If the GCF of the two terms is 1, the expression cannot be factored.

Chapter 6 Answer Key

Are You Ready?—Review
Page 121

1. $m - 10$
2. $w + 12$
3. $\frac{1}{2}g$
4. $\frac{b}{4}$
5. $2a$
6. $s - 8$
7. $m + 50$
8. $7c$

Are You Ready?—Practice
Page 122

1. $n + 15$
2. $s - 9$
3. $\frac{1}{2}p$
4. $c + 7$
5. 50
6. 6
7. 12
8. 28
9. 470 miles
10. 500 visitors

Course 2 • Chapter 6 Equations and Inequalities

Chapter 6 Answer Key

Are You Ready?—Apply
Page 123

1. **STADIUMS** There were 75,000 people in *The Swamp* one Saturday afternoon. The number of the people in *The Swamp* was 1.5 times that of the people in Dolphin Stadium. Solve the equation $1.5p = 75{,}000$ to find the number of people p in Dolphin Stadium.
50,000 people

2. **ALLIGATOR** Linda saw an alligator in the zoo. Dylon told her he had seen an alligator 3 feet longer than the one she saw. The alligator Dylon saw was 12 feet long. Solve the equation $\ell + 3 = 12$ to find the length ℓ of the alligator Linda saw. **9 ft**

3. **FLYING DISK** Marisha threw a flying disk 75 feet. This was 10 feet less than the distance that Quame threw a disk. Solve the equation $d - 10 = 75$ to find the distance d Quame threw the disk.
85 ft

4. **VACATION** Twice as many days as the Jacobi were on vacation was 22. Solve the equation $2v = 22$ to find the number of days v they were on vacation. **11 days**

5. **ENROLLMENT** The number of students in the high school is 300 more than the number of students at the junior high school. There are 900 students at the high school. Solve the equation $900 = j + 300$ to find the number j of students at the junior high.
600 students

6. **WALKING** Tenisha walked 5 blocks to school. The number of blocks Zeke walked to school decreased by 3 was equal to the number of blocks Tenisha walked to school. Solve the equation $z - 3 = 5$ to find the number of blocks z Zeke walked to school. **8 blocks**

Chapter 6 Answer Key

Diagnostic Test
Page 124

1. $p + 14$
2. $s + 3$
3. $n - 12$
4. 14
5. 21
6. 6
7. 12
8. 77
9. about 5,105 m
10. 150 ft

Pretest
Page 125

1. 5
2. -7
3. 9
4. 4
5. -5
6. 21
7. 40
8. $\frac{4}{5}$
9. 1
10. -15
11. $\frac{7}{3}$ or $2\frac{1}{3}$
12. $\frac{4}{13}$
13. $n + 5 \leq 11; n \leq 6$
14. $2n \geq 20; n \geq 10$
15. $x \leq 12;$

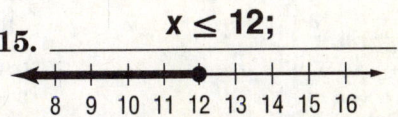

16. $m < -4;$

Course 2 • Chapter 6 Equations and Inequalities

Chapter 6 Answer Key

Chapter Quiz
Page 126

1. −2
2. 12
3. 2
4. 7
5. −4
6. 5
7. 24.4
8. −16
9. $3\frac{1}{3}$
10. 2
11. 11
12. 5.4
13. 5 hours
14. 35 brownies

Vocabulary Test
Page 127

1. equation
2. equivalent equations
3. inequality
4. Addition Property of Inequality
5. two-step equation
6. coefficient
7. Subtraction Property of Equality
8. formula
9. If you multiply each side of an inequality by the same positive number, the inequality stays true.
10. If you multiply each side of an equation by the same number, the two sides remain equivalent.

Chapter 6 Answer Key

Student Recording Sheet, Page 130

Use this recording sheet with the Standardized Test Practice pages.

Fill in the correct answer. For Gridded-Response questions, write your answers in the boxes on the answer grid and fill in the bubbles to match your answers.

1. Ⓐ Ⓑ ● Ⓓ
2. Ⓕ Ⓖ ● Ⓘ
3. Ⓐ Ⓑ Ⓒ ●
4. 6
5. 11

6. ● Ⓖ Ⓗ Ⓘ
7. Ⓐ Ⓑ ● Ⓓ
8. ● Ⓖ Ⓗ Ⓘ
9. $\dfrac{20}{21}$ pizza
10. ● Ⓑ Ⓒ Ⓓ
11. ● Ⓖ Ⓗ Ⓘ
12. 104

13a. $s = (h \div 5) \times 0.5$

13b. 15s; $s = (150 \div 5) \times 0.5 = 15s$

Extended Response
Record your answers for Exercise 13 on the back of this paper.

Chapter 6 Answer Key

Extended-Response Test, Page 131
Sample Answers

In addition to the scoring rubric, the following sample answers may be used as guidance in evaluating extended response assessment items.

1. **a.** The distance d between the two cities is 1,058 miles. The round-trip distance will be twice that. So, the total number of miles t is $2d$; $t = 2,116$ miles.

   b. The Ortiz's car gets 20 miles per gallon. Therefore, the number of gallons g needed is $t \div 20 = g$; $g = 2,116 \div 20$ or 105.8 gallons.

   c. The average price for gas is $2.89 per gallon. The total cost c for gas is $c = \$2.89g$; $c = \$305.76$.

   d. The Ortiz's spend $305.76 on gas, drive 2,116 miles, and know they get 20 miles per gallon gas mileage. Total miles $t \div 20 = g$ (the number of gallons needed). Therefore, $g = 105.8$ gallons. The total cost $c = \$305.76$. Divided by the number of gallons, 105.8, the price per gallon $p = \$305.76 \div 105.8$ gallons; $p = \$2.89$.

2. **a.** $125x + 300 \leq 2,425$

   b. $x \leq 17$; At most, 17 students went on the trip.

Chapter 6 Answer Key

Test, Form 1A
Page 133

1. A
2. G
3. A
4. I
5. C
6. H
7. B
8. F
9. A
10. I
11. A

Test, Form 1A *(continued)*
Page 134

12. I
13. B
14. H
15. C
16. I
17. C
18. G
19. A
20. F

Course 2 • Chapter 6 Equations and Inequalities

Chapter 6 Answer Key

Test, Form 1B
Page 135

1. B
2. F
3. A
4. I
5. B
6. G
7. C
8. G
9. D
10. F
11. C

Test, Form 1B *(continued)*
Page 136

12. F
13. B
14. F
15. A
16. G
17. C
18. I
19. C
20. I

Chapter 6 Answer Key

Test, Form 2A
Page 137

1. B
2. H
3. B
4. F
5. A
6. F
7. D
8. G
9. B
10. H

Test, Form 2A *(continued)*
Page 138

11. D
12. H
13. 170
14. 83.2
15. −16
16. 0.9
17. −8
18. 8.8
19. 9
20. $x \leq 4$;
21. $m > 16$;
22. $p \geq -8$;
23. $w > 5$;
24. $h < -18$;

Course 2 • Chapter 6 Equations and Inequalities

Chapter 6 Answer Key

Test, Form 2B
Page 139

1. B
2. H
3. D
4. I
5. A
6. F
7. D
8. F
9. B
10. G

Test, Form 2B *(continued)*
Page 140

11. A
12. G
13. 240
14. 195.8
15. −30
16. 5
17. −7
18. 64
19. 6
20. $x \leq 8$;
21. $m > 9$;
22. $p \geq -7$;
23. $w > 5$;
24. $h < -12$;

Chapter 6 Answer Key

Test, Form 3A
Page 141

1. −22
2. 12.3
3. 4
4. $300
5. 4
6. −2.5
7. −8
8. −1 1/3
9. −1
10. −2
11. 1.1
12. 20
13. −1
14. −20
15. −4

Test, Form 3A *(continued)*
Page 142

16. −3x = 15
17. x − 5 = 72
18. y > 14;
19. w ≤ 14;
20. k ≥ −9;
21. q > 17;
22. p < −7;
23. b ≤ −9;
24. 3(x − 2) = 27; 11 inches
25. 113 + x < 256; x < 143
26. 0.20x + 4 ≤ 10; x ≤ 30; He can send at most 30 more text messages.

Course 2 • Chapter 6 Equations and Inequalities

Chapter 6 Answer Key

Test, Form 3B
Page 143

1. 12

2. −14

3. $1,075

4. $90

5. 8

6. −5.3

7. −2.6

8. $-\frac{1}{2}$

9. −8.6

10. −4

11. 23

12. 25

13. −2

14. −31

15. −20

Test, Form 3B *(continued)*
Page 144

16. $-6x = -24$

17. $x + 8 = -15$

18. $x > 10$;
    (number line: open circle at 10, arrow right; 8 9 10 11 12)

19. $n \leq 12$;
    (number line: closed circle at 12, arrow left; 9 10 11 12 13)

20. $a \geq -4$;
    (number line: closed circle at −4, arrow right; −6 −5 −4 −3 −2)

21. $m > 4$;
    (number line: open circle at 4, arrow right; 2 3 4 5 6)

22. $w > -16$;
    (number line: open circle at −16, arrow right; −17 −16 −15 −14 −13)

23. $b \leq -12$;
    (number line: closed circle at −12, arrow left; −14 −13 −12 −11 −10)

24. $5(x + 8) = 65$; 5 inches

25. $7x \leq 175$; $x \leq 25$

26. $0.20x + 5 \leq 12$; $x \leq 35$; She can send at most 30 more text messages.

A72   Course 2 • Chapter 6 Equations and Inequalities

Chapter 7 Answer Key

Are You Ready?—Review
Page 145

1. 30°
2. 105°
3. 17.5 in²
4. 40 in²

Are You Ready?—Practice
Page 146

1. 35°
2. 115°
3. 170°
4. 140°
5. 40 ft²
6. 26 in²
7. 45 cm²

Course 2 • Chapter 7 Geometric Figures

Chapter 7 Answer Key

Are You Ready?—Apply
Page 147

1. **GARDENS** Zack is planting a garden that is in the shape of a triangle. The base of the triangle is 12 feet and the height is 11 feet. What is the area of the garden? **66 ft²**	2. **CONSTRUCTION** A builder constructs a deck that contains the angle shown. Use a protractor to find the measure of the angle. **80°** 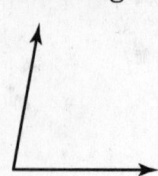
3. **SEWING** Ava cut out a triangle for her quilt. The base of the triangle is 8 inches and the height is 7.5 inches. What is the area of the triangle? **30 in²**	4. **ART** A piece of artwork is made from series of interlocking angles. One of the angles is shown below. Use a protractor to find the measure of the angle. **45°** 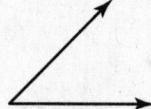
5. **PONDS** A reflecting pond is in the shape of a triangle. It has a base of 21 feet and a height of 15 feet. What is the area of the pond? **157.5 ft²**	6. **ROADS** Two roads intersect to form the angle shown below. Use a protractor to find the measure of the angle. **100°** 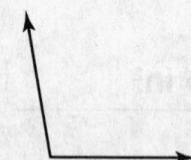

A74

Course 2 • Chapter 7 Geometric Figures

Chapter 7 Answer Key

Diagnostic Test
Page 148

1. 75°
2. 35°
3. 28°
4. 115°
5. 165°
6. 131°
7. 52°
8. 61°
9. 102 orders
10. 4.5 yd²
11. 9 cm²

Pretest
Page 149

1. complementary
2. supplementary
3. obtuse
4. isosceles
5. 84
6. 30
7. 30
8. 16 ft

Course 2 • Chapter 7 Geometric Figures

Chapter 7 Answer Key

Chapter Quiz
Page 150

1. Sample answer: ∠2 and ∠4
2. obtuse
3. right
4. supplementary
5. complementary
6. acute; 45
7. acute; 45
8. 10 ways

Vocabulary Test
Page 151

1. complementary angles
2. obtuse angle
3. triangle
4. scale
5. pyramid
6. congruent segments
7. scale factor
8. plane

Chapter 7 Answer Key

Student Recording Sheet, Page 154
Use this recording sheet with the Standardized Test Practice pages.

Fill in the correct answer. For gridded-response questions, write your answers in the boxes on the answer grid and fill in the bubbles to match your answers.

1. Ⓐ ● Ⓒ Ⓓ

2. ● Ⓖ Ⓗ Ⓘ

3.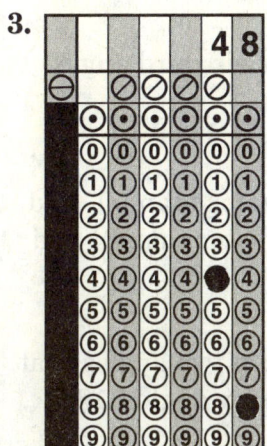

4. Ⓐ ● Ⓒ Ⓓ

5. ● Ⓖ Ⓗ Ⓘ

6. ● Ⓑ Ⓒ Ⓓ

7. right triangle

8. Ⓕ ● Ⓗ Ⓘ

9. 50°

10. Ⓐ Ⓑ Ⓒ ●

11. Ⓕ ● Ⓗ Ⓘ

12. Ⓐ ● Ⓒ Ⓓ

13. Ⓕ Ⓖ ● Ⓘ

14a. obtuse

14b. acute

14c. obtuse isosceles

14d. 26°; The sum of the measures of a triangle must be 180°. $128 + 26 + 26 = 180$

Extended Response
Record your answers for Exercise 14 on the back of this paper.

Chapter 7 Answer Key
Extended-Response Test, Page 155
Sample Answers

In addition to the scoring rubric, the following sample answers may be used as guidance in evaluating extended response assessment items.

1. **a.** If two angles have the same measure, they are congruent.

   b. $\angle 1 \cong \angle 3$;
   $\angle 2 \cong \angle 4$;
   $m\angle 1 = 104°$;
   $m\angle 3 = 104°$;
   $m\angle 4 = 76°$

2. **a.** All triangles have two acute angles. Triangles are classified by their third angle.

   See students' triangles.

   b. See students' angles. Right angles are 90°; obtuse angles are greater than 90° and less than 180°; straight angles are 180°, and acute angles are greater than 0° and less than 90°.

   c. The sum of two angles that are supplementary is 180°. The sum of two angles that are complementary is 90°, which is half the total for supplementary angles, 180°.

   See students' drawings.

3. **a.** The figure has two triangular bases and three rectangular faces. The figure is a triangular prism.

   b. Triangles *ABC* and *DEF*; The bases are parallel. The triangles are the only parallel figures in the figure.

Chapter 7 Answer Key

Test, Form 1A
Page 157

1. A
2. H
3. C
4. I
5. B
6. I

Test, Form 1A *(continued)*
Page 158

7. A
8. I
9. C
10. H
11. C
12. F

Course 2 • Chapter 7 Geometric Figures

A79

Chapter 7 Answer Key

Test, Form 1B
Page 159

1. _____D_____

2. _____H_____

3. _____A_____

4. _____H_____

5. _____A_____

6. _____H_____

Test, Form 1B *(continued)*
Page 160

7. _____B_____

8. _____F_____

9. _____C_____

10. _____H_____

11. _____A_____

11. _____H_____

Chapter 7 Answer Key

Test, Form 2A
Page 161

1. A

2. I

3. D

4. H

5. A

6. I

Test, Form 2A *(continued)*
Page 162

7. $6\frac{3}{4}$ in. by 5 in.

8. 40

9.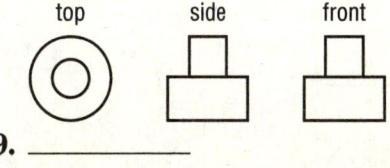

10. right scalene

11. hexagonal pyramid

12. triangle

Chapter 7 Answer Key

Test, Form 2B
Page 163

1. D

2. I

3. B

4. H

5. D

6. I

Test, Form 2B (continued)
Page 164

7. $4\frac{1}{2}$ in. by 4 in.

8. 130

9.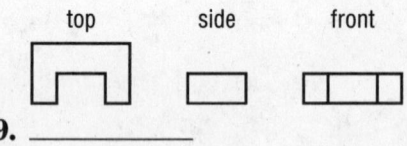

10. acute equilateral

11. triangular prism

12. trianlge

A82

Course 2 • Chapter 7 Geometric Figures

Chapter 7 Answer Key

Test, Form 3A
Page 165

1. 28

2. 11

3. 70

4. 15

5a. $2\frac{1}{3}$ in; $\frac{1}{72}$

5b. 32 cm; $\frac{1}{125}$

6. 63°; acute

7. 90°; right

8. 210 ft

Test, Form 3A *(continued)*
Page 166

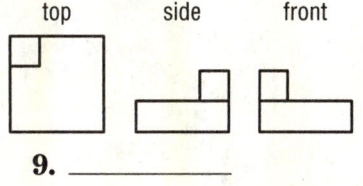

9.

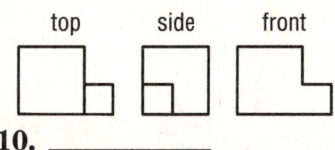

10.

11. Figure name: square pyramid
base: ABCD
faces: ABCD, ABE, BCE, CDE, ADE
edges: AE, BE, CE, DE, AB, BC, CD, AD
vertices: A, B, C, D, E

12. hexagon

13. If the cross section is an angled slice it will be a trapezoid.

Course 2 • Chapter 7 Geometric Figures

A83

Chapter 7 Answer Key

Test, Form 3B
Page 167

1. 3
2. 20
3. 55
4. 30
5a. $29\frac{1}{11}$ cm; $\frac{1}{1,650}$
5b. 18 in.; $\frac{1}{4}$
6. 53°; right
7. 120°; obtuse
8. 312.5 ft

Test, Form 3B *(continued)*
Page 168

9.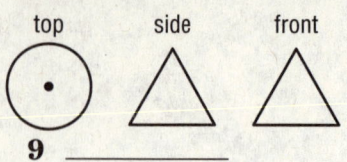
10.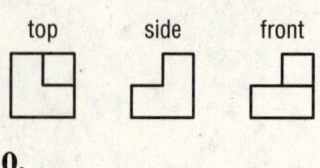
11. Figure name: pentagonal pyramid
base: ABCDE
faces: ABF, BCF, CDF, DEF, AEF, ABCDE
edges: AB, BC, CD, DE, EA, AF, BF, CF, DF, EF
vertices: A, B, C, D, E, F
12. rectangle
13. Each cross section would be a triangle.

Chapter 8 Answer Key

Are You Ready?—Review
Page 169

1. 84 mm²
2. 14 m²
3. 60 yd²
4. 8.04 cm²
5. 93.5 in²

Are You Ready?—Practice
Page 170

1. 24.42 cm²
2. 43.68 ft²
3. 24.5 mm²
4. 119 in²
5. 38 m²
6. 71.5 yd²
7. 287.5 in²
8. 6.9 yd²

Course 2 • Chapter 8 Measure Figures

Chapter 8 Answer Key

Are You Ready?—Apply
Page 171

1. **FARMING** A farmer has a rectangular field that is 100 yards long by 20 yards wide. What is the total area of the field?
2,000 yd²

2. **WALLPAPER** Sydney is going to wallpaper the rectangular wall in her bedroom. If her wall is 16 feet long by 8 feet tall, how much wallpaper will she need?
128 ft²

3. **GARDENING** Elena wants to put a new layer of soil in her triangular garden. She needs to know the area of her garden so that she knows how much soil to buy. If the garden has a base of 14.3 meters and a height of 6 meters, how much area does she need to cover?
42.9 m²

4. **TILES** Mr. McCabe is buying tiles for his rectangular kitchen floor. The floor is 15 feet long by 25 feet wide. If each tile is 1 square foot, how many tiles does he need?
375 tiles

5. **ARTS** Ariel is making a sign in the shape of a triangle. The triangle has a base of 30.5 inches and a height of 36 inches. What is the area of the triangle?
549 in²

6. **WILDFIRES** A spokesperson from the forest fire service said that a wildfire caused a rectangular patch of forest that was 26 miles long by 24 miles wide to be burned. How many square miles of forest were burned by the wildfire?
624 mi²

Chapter 8 Answer Key

Diagnostic Test
Page 172

1. _____ 61.32 cm² _____

2. _____ 14.08 ft² _____

3. _____ 37.72 m² _____

4. _____ 21.39 m² _____

5. _____ 165 mm² _____

6. _____ 875 in² _____

7. _____ 88 in² _____

8. _____ 3,200 ft² _____

Pretest
Page 173

1. _____ 864 in³ _____

2. _____ 240 cm³ _____

3. _____ 1,000 in³ _____

4. _____ 120 mm³ _____

5. _____ 108 cm² _____

6. _____ 448 mm² _____

Chapter 8 Answer Key

Chapter Quiz
Page 174

1. 15.7 in.

2. 154 cm

3. 154 yd^2

4. 28.3 m^2

5. 88.3 ft^2

6. 518.4 ft^3

7. 13,888 in^3

Vocabulary Test
Page 175

1. i
2. f
3. h
4. c
5. b
6. g
7. e
8. d
9. a
10. g
11. sum of the areas of all the lateral faces

Chapter 8 Answer Key

Student Recording Sheet, Page 178
Use this recording sheet with the Standardized Test Practice pages.

Fill in the correct answer. For gridded-response questions, write your answers in the boxes on the answer grid and fill in the bubbles to match your answers.

1. Ⓐ Ⓑ Ⓒ ●
2. Ⓕ Ⓖ ● Ⓘ
3. 1/8
4. 95/3

5. Ⓐ Ⓑ ● Ⓓ
6. Ⓕ Ⓖ ● Ⓘ
7. −3

8. Ⓐ ● Ⓒ Ⓓ
9. The surface area of the second pyramid is greater. 129 cm² > 96 cm²
10. Ⓕ Ⓖ ● Ⓘ
11. 12.56 in²
12. Ⓐ ● Ⓒ Ⓓ
13a. 216 cm³
13b. 6,184 cm³
13c. 2,080 cm³

Extended Response
Record your answers for Exercise 13 on the back of this paper.

Course 2 • Chapter 8 Measure Figures

Chapter 8 Answer Key

Extended-Response Test, Page 179
Sample Answers

In addition to the scoring rubric, the following sample answers may be used as guidance in evaluating extended response assessment items.

1. **a.** See students' drawings.

   b. The diameter can be found by multiplying the radius by 2. The circumference can be found by using the formula $C = \pi d$ or $C = 2\pi r$. The diameter is 2×7, or 14 m. Using $C = \pi d$, the circumference is about 3×14 or 42 m. Or using $C = 2\pi r$, the circumference is about $2 \times 3 \times 7$, or 42 m.

2. **a.** Volume of A: 72 in^3
   Volume of B: 144 in^3
   Volume of C: 144 in^3
   $2 \times$ Volume A = Volume B = Volume C
   Surface Area of A: 120 in^2
   Surface Area of B: 168 in^2
   Surface Area of C: 216 in^2
   Surface Area of A + 48 = Surface Area of B
   Surface Area of B + 48 = Surface Area of C

   b. An increase in height gives the appearance of a larger increase in surface area and volume than does an increase in width, depth, or radius.

Chapter 8 Answer Key

Test, Form 1A
Page 181

1. C
2. G
3. C
4. F
5. B
6. H
7. B
8. H

Test, Form 1A *(continued)*
Page 182

9. C
10. I
11. B
12. F
13. A

Chapter 8 Answer Key

Test, Form 1B
Page 183

1. D
2. H
3. B
4. H
5. C
6. F
7. D
8. G

Test, Form 1B *(continued)*
Page 184

9. C
10. H
11. B
12. G
13. D

Chapter 8 Answer Key

Test, Form 2A
Page 185

1. C

2. H

3. B

4. H

5. C

6. I

7. C

Test, Form 2A *(continued)*
Page 186

8. F

9. 164 ft²; 80 ft³

10. 266.7 m³

11. 61.3 ft²

12. 1,156.7 cm²

Course 2 • Chapter 8 Measure Figures

Chapter 8 Answer Key

Test, Form 2B
Page 187

1. B

2. G

3. D

4. I

5. C

6. G

7. C

Test, Form 2B *(continued)*
Page 188

8. H

9. 184 ft²; 99 ft³

10. 74.7 m³

11. 54.7 ft²

12. 849.9 cm²

Chapter 8 Answer Key

Test, Form 3A
Page 189

1. 76.7 in³

2. 600 m³

3. 810 ft³

4. 619.7 m³

5. 36.4 cm

6. 907.5 in²

7. 46.2

Test, Form 3A *(continued)*
Page 190

8. 294 m²

9. 64 ft³

10. 105 in²

11. 480 m³

12. 384 m²

13. 14 in.

14. Sample answer: base area equals 57 cm² and height equals 10 cm; base area equals 38 cm² and height equals 15 cm

Chapter 8 Answer Key

Test, Form 3B
Page 191

1. 6,335.6 mm³

2. 715 m³

3. 486 ft³

4. 1,252.3 m³

5. 141.3 ft

6. 2,550.5 m²

7. 40.8

Test, Form 3B *(continued)*
Page 192

8. 486 m²

9. 50 ft³

10. 144 in²

11. 13,500 m³

12. 3,672 m²

13. $2\frac{1}{4}$ ft

14. Sample answer: base area equals 63 cm² and height equals 10 cm; base area equals 30 cm² and height equals 21 cm

Chapter 9 Answer Key

Are You Ready?—Review
Page 193

1. $\dfrac{1}{9}$
2. $\dfrac{2}{3}$
3. $\dfrac{2}{3}$
4. $\dfrac{1}{3}$
5. $\dfrac{2}{7}$
6. simplified
7. $\dfrac{3}{8}$
8. $\dfrac{3}{5}$
9. simplified
10. $\dfrac{7}{10}$
11. $\dfrac{3}{10}$
12. $\dfrac{3}{10}$

Are You Ready?—Practice
Page 194

1. 126
2. 180
3. 336
4. 207
5. 840
6. 1,320
7. 504
8. 120
9. 7
10. 10
11. 30
12. 2
13. $\dfrac{1}{4}$
14. $\dfrac{1}{6}$
15. simplified
16. $\dfrac{1}{3}$
17. $\dfrac{2}{3}$
18. $\dfrac{1}{5}$

Chapter 9 Answer Key

Are You Ready?—Apply
Page 195

Solve.

1. Marjorie ran 3 miles in 21 minutes. Write a fraction, in simplest form, that represents the unit rate in minutes per mile. $\dfrac{7}{1}$

2. The average person sleeps 8 hours a day. About how many hours does the average person sleep during a lifetime of 80 years? Use 365 as the number of days in one year. **233,600 hours**

3. The table shows the number of students in Mrs. Blair's math class that prefer each sport. There are 26 students in the class. What fraction of the class, in simplest form, prefers basketball? $\dfrac{4}{13}$

Preferred Sport	
Sport	Students
Baseball	4
Basketball	8
Football	12
Other	2

4. Karen has eight CDs in her music collection. Each CD has nine songs on it. If each song lasts an average of 4 minutes, about how long could Karen play all of her CDs without repeating a song? **288 minutes**

5. Hector makes $9 per hour mowing lawns. If he mows lawns for 4 hours each day, how much will he earn after 8 days? **$288**

6. A 30-minute television show contained a total of about 6 minutes of commercials. What fraction, in simplest form, of the 30 minutes was devoted to commercials? $\dfrac{1}{5}$

7. Leonard baked six batches of chocolate chip cookies. Each batch had 12 cookies. Each cookie had an average of 6 chocolate chips in it. About how many chocolate chips were in the six batches of cookies? **432**

8. Jasmine recorded the number of hours she spent doing each activity one day. The table shows the results. What fraction of her day, in simplest form, was spent online? $\dfrac{1}{6}$

Daily Activities	
Activity	Hours
Eating	2
Online	4
Reading	3
Sleeping	9
With Friends	6

Chapter 9 Answer Key

Diagnostic Test
Page 196

1. 198
2. 68
3. 186
4. 357
5. 1,680
6. 2,184
7. 60
8. 7,920
9. 6
10. 20
11. 22
12. 5
13. $\frac{1}{7}$
14. $\frac{2}{11}$
15. $\frac{1}{2}$
16. simplified
17. $\frac{1}{8}$
18. $\frac{2}{3}$

Pretest
Page 197

1. 12
2. 18
3. 8
4. $\frac{1}{2}$
5. $\frac{2}{3}$
6. $\frac{1}{3}$
7. $\frac{5}{6}$
8. 4,896
9. 120
10. 90
11. independent
12. dependent
13. independent

Chapter 9 Answer Key

Chapter Quiz
Page 198

1. $\dfrac{1}{3}$
2. $\dfrac{2}{3}$
3. $\dfrac{1}{2}$
4. 6
5. $\dfrac{1}{4}$
6. $\dfrac{13}{50}$
7. $\dfrac{1}{4} < \dfrac{7}{25}$
8. 1, 1; 1, 2; 1, 3; 1, 4; 2, 1; 2, 2; 2, 3; 2, 4; 3, 1; 3, 2; 3, 3; 3, 4; 4, 1; 4, 2; 4, 3; 4, 4; 5, 1; 5, 2; 5, 3; 5, 4; 6, 1; 6, 2; 6, 3; 6, 4; $\dfrac{3}{4}$
9. Sample answer: flip a coin 25 times

Vocabulary Test
Page 199

1. independent events
2. permutation
3. Fundamental Counting Principle
4. probability
5. simple
6. complementary events
7. sample space
8. Experimental probability
9. tree diagrams
10. Sample answer: a possible result of an experiment
11. Sample answer: an experiment designed to act out a given situation

Chapter 9 Answer Key

Student Recording Sheet, Page 202

Use this recording sheet with the Standardized Test Practice pages.

Fill in the correct answer. For gridded-response questions, write your answers in the boxes on the answer grid and fill in the bubbles to match your answers.

1. A ● C D

2. F G H ●

3. A ● C D

4. 5/24

5. 360

6. F ● H I

7. A B ● D

8. point *A*

9. ● G H I

10. BCEM, BCME, EBCM, EMBC, MBCE, and MEBC

11. A B C ●

12. F ● H I

13a. 4; 47N − 27W; 129N − 27W; 26W − 19W − 349N − 27W; 26W − 19W − 51N

13b. $\frac{3}{4}$

Extended Response

Record your answers for Exercise 13 on the back of this paper.

Chapter 9 Answer Key

Extended-Response Test, Page 203
Sample Answers

In addition to the scoring rubric, the following sample answers may be used as guidance in evaluating extended response assessment items.

1. **a.** There are 4 possible outcomes having an equal chance of occurring.

   $P(1) = \dfrac{1}{4}$ ← number of ways to spin a 1
   ← number of possible outcomes

   b. Spin spinner B 100 times and tally the number of times you spin W.

   $P(W) = \dfrac{n}{100}$ ← number of times you spin W

2. **a.**

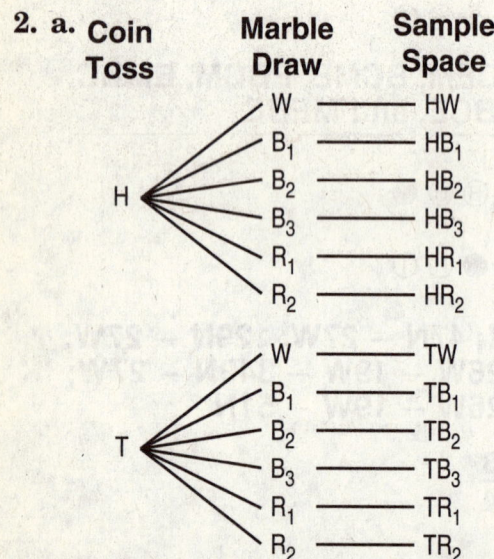

   b. If the outcome of one event does not influence the outcome of a second event, the events are called independent events.

   c. To find the probability of tossing a head and drawing a red marble, write the number of ways to toss a head and draw a red marble (2) over the number of possible outcomes.
   $P(H, R) = P(H) \cdot P(R)$
   $= \dfrac{1}{2} \cdot \dfrac{2}{6}$ or $\dfrac{1}{6}$, since they are independent events.

   d. $P(B, \text{then } B) = P(B) \cdot P(B \text{ after } B)$
   $= \dfrac{1}{2} \cdot \dfrac{2}{5}$ or $\dfrac{1}{5}$, The probability of the first event is $\dfrac{1}{2}$ because $\dfrac{1}{2}$ of the marbles are blue. The probability of the second event is $\dfrac{2}{5}$ because the blue marble drawn was not replaced. Two of the remaining five marbles are blue.

3. **a.** Each time Rayna chooses one of each type of relative is an event. The event of choosing a niece can occur in 8 ways, choosing a nephew can occur in 9 ways, and so on. So, Rayna could choose a possible $8 \times 9 \times 6 \times 3 \times 5$ or 6,480 teams.

   b. $8 \cdot 7 \cdot 6 \cdot 5 \cdot 4 \cdot 3 \cdot 2 \cdot 1$ or 40,320; The order of the arrangement of the nieces is important, so a permutation is used.

   c. The event of choosing a niece can occur in 8 ways, choosing a nephew can occur in 9 ways, and choosing a cousin can occur in 6 ways. So Rayna could choose a possible $8 \times 9 \times 6$ or 432 teams.

Chapter 9 Answer Key

Test, Form 1A
Page 205

1. B
2. G
3. B
4. I
5. D
6. H
7. D
8. G

Test, Form 1A *(continued)*
Page 206

9. B
10. G
11. C
12. H
13. C
14. I
15. D
16. F
17. C

Course 2 • Chapter 9 Probability

Chapter 9 Answer Key

Test, Form 1B
Page 207

1. A
2. G
3. D
4. H
5. D
6. I
7. A
8. I

Test, Form 1B *(continued)*
Page 208

9. B
10. F
11. C
12. G
13. D
14. H
15. A
16. H
17. B

Chapter 9 Answer Key

Test, Form 2A
Page 209

1. C
2. F
3. A
4. F
5. B
6. G
7. C

Test, Form 2A *(continued)*
Page 210

8. F
9. B
10. H
11. $\dfrac{1}{3}$
12. $\dfrac{2}{3}$
13. 24 ways
14. $\dfrac{4}{207}$
15. 95,040

Course 2 • Chapter 9 Probability

Chapter 9 Answer Key

Test, Form 2B
Page 211

1. C

2. I

3. D

4. H

5. C

6. F

7. B

Test, Form 2B *(continued)*
Page 212

8. F

9. B

10. F

11. $\dfrac{1}{3}$

12. $\dfrac{5}{6}$

13. 6 ways

14. $\dfrac{1}{36}$

15. 720

Chapter 9 Answer Key

Test, Form 3A
Page 213

1. 120 orders

2. 720 codes

3. 650 ways

4. Theor. prob. $\left(\frac{1}{5}\right)$ < exp. prob. $\left(\frac{2}{5}\right)$

5. Sample answer: Use a random number function of a graphing calculator.

6. 1,512 outcomes

7. 96 outcomes

8. 24 ways

Test, Form 3A *(continued)*
Page 214

9. $\frac{1}{2}$

10. $\frac{1}{4}$

11. $\frac{7}{8}$

12. $\frac{1}{64}$

13. $\frac{1}{19}$

14. $\frac{6}{95}$

15. Theor. prob. $\left(\frac{1}{6}\right)$ < exp. prob. $\left(\frac{1}{5}\right)$

16. 6,720

17. 90

18. 7,920

19. $\frac{3}{95}$

20. $\frac{25}{99}$; dependent event; The second event is impacted by the first.

Course 2 • Chapter 9 Probability

Chapter 9 Answer Key

Test, Form 3B
Page 215

1. 24 orders

2. 30,240 codes

3. 506 ways

4. Theor. prob. $\left(\frac{1}{5}\right)$ < exp. prob. $\left(\frac{4}{17}\right)$

5. Sample answer: Use a random number function of a graphing calculator.

6. 672 outcomes

7. 48 outcomes

8. 120 ways

Test, Form 3B *(continued)*
Page 216

9. $\frac{4}{7}$

10. $\frac{6}{7}$

11. $\frac{2}{7}$

12. $\frac{1}{49}$

13. $\frac{3}{38}$

14. $\frac{5}{76}$

15. Theor. prob. $\left(\frac{1}{6}\right)$ > exp. prob. $\left(\frac{1}{7}\right)$

16. 24

17. 120

18. 15,120

19. $\frac{7}{100}$

20. $\frac{2}{65}$; dependent event; The second event is impacted by the first.

Chapter 10 Answer Key

Are You Ready?—Review
Page 217

1. 98; 74

2. 50%

3. 55 people

4. 64 people

5. 204 students

Are You Ready?—Practice
Page 218

1. Sophia

2. Ryan, Charlotte

3. 75°F; 83°F

4. 25%

5. 56 students

Chapter 10 Answer Key

Are You Ready?—Apply
Page 219

1. **SOCCER** The bar graph shows the number of years that five players have been playing soccer. Which player has been playing soccer the longest? **Ella**

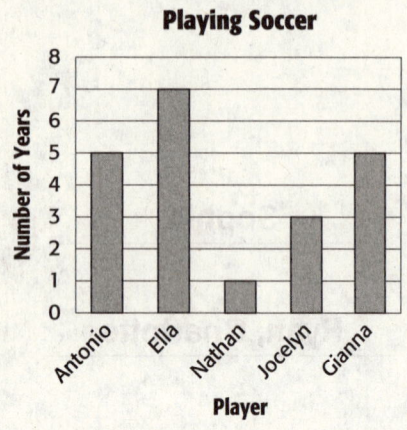

2. **INTERNET** The circle graph shows the results of a survey about the number of days each week that students go online. If 300 students were surveyed, how many students go online 6 or 7 days each week?
105 students

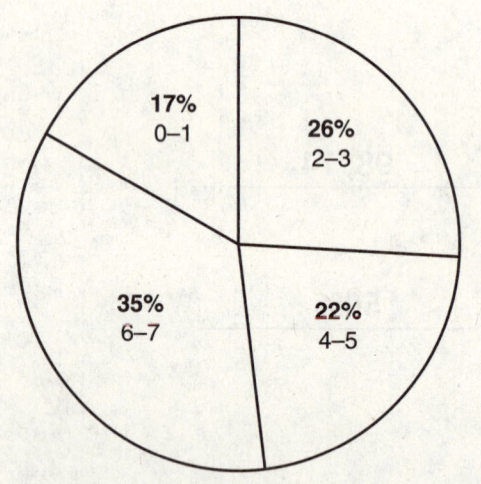

3. **SOCCER** Refer to the bar graph in Exercise 1. Which player(s) have been playing soccer for less than 4 years?
Nathan, Jocelyn

4. **SURVEYS** The Booster Club distributed a survey to 400 families. Of those who received the survey, 76% returned the survey. How many families returned the survey?
304 families

5. **MOVIES** In a survey, 220 people were asked to describe the average number of times in a month that they go to a movie theater. Of those surveyed, 45% said they go to the movies more than 3 times a month. How many people surveyed go to the movies more than 3 times a month? **99 people**

6. **GOLF** The table shows the miniature golf scores of twenty people. What percent of the people had scores of 50 or less? **25%**

Miniature Golf Scores				
54	56	50	48	71
63	62	45	51	50
55	65	60	53	70
64	60	55	50	52

Chapter 10 Answer Key

Diagnostic Test
Page 220

1. blue shark
2. night shark
3. 40 in.; 68 in.
4. 25%
5. 78 students

Pretest
Page 221

1. Sample answer: bar graph
2. Sample answer: line graph
3. Sample answer: bar graph
4. Sample answer: stem-and-leaf plot
5. 9 students
6. Sample answer: The vertical scale jumps from 0 to 400. This makes the luggage lost for Company A look like 4 times that of Company B.

Chapter 10 Answer Key

Chapter Quiz
Page 222

1. 57 students

2. Mean: $1,110; Median: $1,200; Mode: $1,200; The mean might be misleading because the mean price is lower than the majority of the prices.

3. about 761

4. about 307 students

5. Valid, it is an unbiased systematic random sample.

6. Not valid, it is a biased voluntary response sample.

Vocabulary Test
Page 223

1. unbiased sample

2. population

3. survey

4. biased survey

5. voluntary response sample

6. statistics

7. double box plot

8. Sample answer: Every member of the population has the same chance of being chosen.

9. Sample answer: Part of the population that is being surveyed.

Chapter 10 Answer Key

Student Recording Sheet, Page 226
Use this recording sheet with the Standardized Test Practice pages.

Fill in the correct answer. For gridded-response questions, write your answers in the boxes on the answer grid and fill in the bubbles to match your answers.

1. Ⓐ Ⓑ Ⓒ ●
2. Ⓕ Ⓖ Ⓗ ●
3. Ⓐ Ⓑ Ⓒ ●
4. 2.3
5. 5/2

6. Ⓕ Ⓖ ● Ⓘ
7. ● Ⓑ Ⓒ Ⓓ
8. Ⓕ Ⓖ ● Ⓘ
9. ● Ⓑ Ⓒ Ⓓ
10. $\frac{5}{20} = \frac{x}{100}$; 25%

11a. Line graphs show change over time.

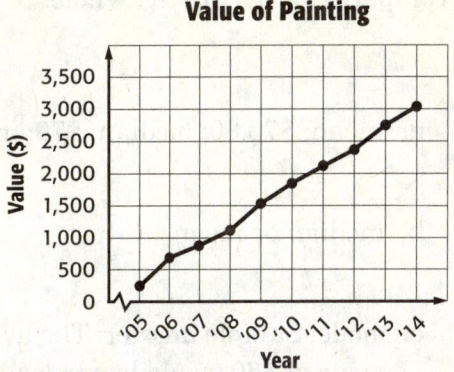

11b. Sample answer: The value of the painting increased from 2005 to 2014.

11c. Sample answer: $4,000

Extended Response
Record your answers for Exercise 11 on the back of this paper.

Chapter 10 Answer Key

Extended-Response Test, Page 227
Sample Answers

In addition to the scoring rubric, the following sample answers may be used as guidance in evaluating extended response assessment items.

1. **a.** Sample answer: line graph; the data is given over a period of time

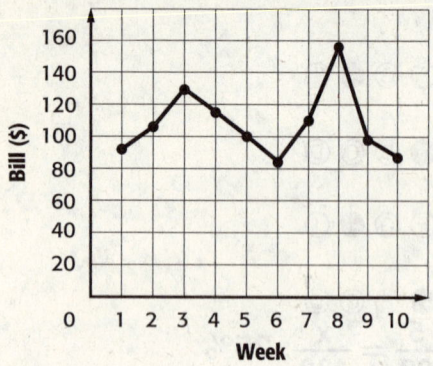

2. Sample answer: An appropriate display would be a circle graph because you are comparing a part to the whole.

3. **a.** mean; $70.80; median: $70; mode: $60

   b. median or mean

   c. mode; Sample answer: The average price for a pair of sunglasses is $70.80. The mode of $60 would be misleading since it is much lower than the average price.

Chapter 10 Answer Key

Test, Form 1A
Page 229

Test, Form 1A *(continued)*
Page 230

1. D

2. F

3. C

4. I

5. B

6. H

7. C

8. I

9. A

10. H

Course 2 • Chapter 10 Statistics

A115

Chapter 10 Answer Key

Test, Form 1B
Page 231

1. A

2. G

3. D

4. H

5. B

6. I

Test, Form 1B *(continued)*
Page 232

7. C

8. H

9. A

10. F

Chapter 10 Answer Key

Test, Form 2A
Page 233

1. __A__

2. __I__

3. __A__

4. __G__

Test, Form 2A *(continued)*
Page 234

5. __A__

6. Sample answer: The teacher used the mean of 6.5, which is greater than the median, which is 5.5, and mode which is 5.

7. $\frac{4}{5}$ of the team is actually shorter than 72 inches.

8. valid; This is an unbiased random sample of renters.

Chapter 10 Answer Key

Test, Form 2B
Page 235

Test, Form 2B *(continued)*
Page 236

1. B
2. G
3. A
4. I

5. D

6. Sample answer: The principal used the mean of 26, which is greater than the median, 23.5, and the mode, 16.

7. valid; This is an unbiased systematic random sample of park visitors.

8. mean

Chapter 10 Answer Key

Test, Form 3A
Page 237

1. Sample answer: line graph;
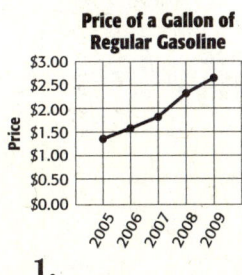

2. Room 240

3. Sample answer: Room 300 had higher test scores.

4. 24.75; 26; no mode

5. median; It is the greatest value.

6. Sample answer: histogram;
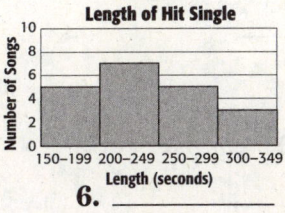

Test, Form 3A *(continued)*
Page 238

7. Sample answer: An unbiased sample is representative of the entire population so it gives valid results.

8. 157,500 voters

9. 291 students

10. Invalid; It is based on a biased voluntary response sample.

11. Valid; It is based on an unbiased simple random sample.

12. mean absolute deviation

Chapter 10 Answer Key

Test, Form 3B
Page 239

1. Sample answer: line graph;

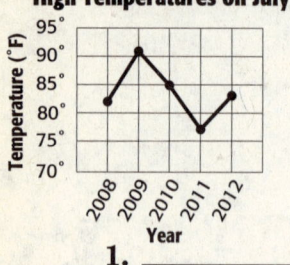

2. Room 12

3. Sample answer: Room 14 had higher test scores.

4. 20, 15, 12 and 15

5. The principal used the mean which is 20. This is much greater than the median of 15 and modes of 12 and 15.

6. Sample answer: histogram;

Test, Form 3B *(continued)*
Page 240

7. simple random sample and systematic random sample; convenience sample and voluntary response sample

8. 12,500 voters

9. 172 students

10. Valid; It is based on an unbiased systematic random sample.

11. Invalid; It is based on a biased convenience sample.

12. interquartile range

Benchmark Test Answer Keys

Course 2 Benchmark Test – First Quarter

1. The table shows the costs of different size jars of peanut butter. Which of the jars has the lowest unit rate?

Comparison Shopping	
Size	Cost
12-oz	$3.00
18-oz	$4.40
25-oz	$6.75
32-oz	$8.25

A. 12-oz jar
*B. 18-oz jar
C. 25-oz jar
D. 32-oz jar

2. The enrollment at a community college this year is 115% of last year's enrollment. If there were 1,240 students enrolled at the college last year, how many students are there this year?

F. 1,054 students
G. 1,302 students
H. 1,378 students
*I. 1,426 students

3. Vicky jogged $2\frac{3}{4}$ miles in $\frac{1}{2}$ hour. What was her average rate of speed in miles per hour?

A. $1\frac{3}{8}$ miles per hour
B. $3\frac{1}{4}$ miles per hour
*C. $5\frac{1}{2}$ miles per hour
D. $6\frac{3}{4}$ miles per hour

4. In a recent survey, 55% of pet owners have more than one pet. If there were 620 pet owners surveyed, which proportion can be used to find the number who own more than one pet?

F. $\frac{100}{55} = \frac{n}{620}$
*G. $\frac{55}{100} = \frac{n}{620}$
H. $\frac{55}{100} = \frac{620}{n}$
I. $\frac{55}{620} = \frac{n}{100}$

5. SHORT ANSWER A pair of jeans that normally sells for $35 is on sale for 20% off. Find the sale price of the jeans. Then find the total cost of the jeans if the sales tax rate is 6%.

$28; $29.68

6. How much simple interest is earned on an investment of $1,250 if the money is invested for 5 years at an annual interest rate of 4.5%?

A. $1,531.25
B. $1,306.25
*C. $281.25
D. $56.25

Course 2 Benchmark Test – First Quarter (continued)

7. SHORT ANSWER Determine whether the relationship between the two quantities in the table is proportional. Explain your reasoning.

Bicycle Rental	
Hours	Cost ($)
0	12.50
1	17.50
2	22.50
3	27.50

The relationship is not proportional. The line formed by the ordered pairs (hours, cost) does not intersect the origin.

8. What is the slope of the line that passes through points R and T?

F. $-\frac{3}{1}$
G. $-\frac{1}{3}$
H. $\frac{1}{3}$
*I. $\frac{3}{1}$

9. The weight of an object on the moon varies directly as the weight of the object on Earth. A 90-pound object on Earth weighs 15 pounds on the moon. If an object weighs 156 pounds on Earth, how much does it weigh on the moon?

A. 23 pounds
*B. 26 pounds
C. 28 pounds
D. 936 pounds

10. A muffin recipe calls for 4 cups of sugar and yields 36 muffins. If Amelia only wants to make 24 muffins, how much sugar will she need?

F. 6 cups
G. $3\frac{3}{4}$ cups
*H. $2\frac{2}{3}$ cups
I. $2\frac{1}{2}$ cups

11. A sprinter runs 100 meters in 11.5 seconds. What is the runner's average running rate in meters per second? Round to the nearest tenth.

*A. 8.7 meters per second
B. 9.5 meters per second
C. 10.1 meters per second
D. 11.5 meters per second

Benchmark Test Answer Keys

Course 2 Benchmark Test – First Quarter (continued)

12. Amy earns $7 per hour for babysitting. Which of the following statements is true about the relationship between the number of hours Amy works and her earnings?

   *F. The relationship is proportional because the graph of the line passes through the origin and has a constant rate of change.

   G. The relationship is proportional because there is not a constant rate of change between the points.

   H. The relationship is nonproportional because the points do not form a straight line.

   I. The relationship is nonproportional because the line through the points does not intersect the origin.

13. A video game that normally sells for $80 is on sale for $68. What is the percent of discount for the sale price?

   A. 18%
   B. 17%
   *C. 15%
   D. 12%

14. What is the constant rate of change of the ordered pairs shown in the table?

x	y
2	3
4	7
6	11
8	15

   F. 1
   *G. 2
   H. 3
   I. 4

15. **SHORT ANSWER** Estimate 58% of 121 by using 10%. Show your work.

   Sample answer: about $72; Round 121 to 120. Round 58% to 60%. 10% of 120 is 12 and 6 times 12 is 72.

16. Last year there were 43 science projects submitted by students at a science fair. This year there are 52 science projects. To the nearest tenth, what is the percent of change in the number of science projects submitted?

   A. 17.3% decrease
   B. 17.3% increase
   C. 20.9% decrease
   *D. 20.9% increase

Course 2 Benchmark Test – First Quarter (continued)

17. Simplify the complex fraction.

   $$\frac{\frac{4}{3}}{\frac{2}{5}}$$

   F. $\frac{3}{10}$
   G. $\frac{8}{15}$
   H. $\frac{15}{8}$
   *I. $\frac{10}{3}$

18. What percent of the figure below is shaded?

   *A. 45%
   B. 40%
   C. 20%
   D. 18%

19. Mr. Thompson plans to invest $7,500 in a savings account that earns 2.75% simple annual interest. If he makes no other deposits or withdrawals, how much money will Mr. Thompson's account be worth after 10 years?

   F. $2,062.50
   G. $7,706.25
   *H. $9,562.50
   I. $10,128.25

20. **SHORT ANSWER** Use the percent equation to solve the following problem. Show your work.

   98 is 35% of what number?

   **280; 98 = 0.35 · w;
   98 ÷ 0.345 = w; w = 280**

21. What is the slope of the line shown on the coordinate plane?

   *A. 1
   B. $\frac{3}{5}$
   C. $-\frac{3}{5}$
   D. -1

22. Which of the following equations represents a direct variation?

   F. $y = x - 1$
   *G. $y = \frac{x}{3}$
   H. $y = x + 5$
   I. $y = 2x - 3$

Benchmark Test Answer Keys

Course 2 Benchmark Test – First Quarter (continued)

23. The bookstore normally sells mechanical pencils for $6.50. This week the pencils are discounted by 25%. To the nearest cent, what is the amount of discount?

    A. $1.30
    *B. $1.63
    C. $2.11
    D. $4.88

24. Christy drove 135 miles in 2.5 hours. What was her average speed in miles per hour?

    F. 50 miles per hour
    G. 52 miles per hour
    *H. 54 miles per hour
    I. 55 miles per hour

25. **SHORT ANSWER** An electrician charges a $50 fee to make a service call plus $25 per hour he works. Complete the table. Then determine whether the relationship between the two variables is proportional. Explain your reasoning.

    Cost of Hiring an Electrician

Hours	Cost ($)
1	75
2	100
3	125
4	150

    The relationship is not proportional because the graph of the relationship does not pass through the origin.

Course 2 Benchmark Test – Second Quarter

1. Which two points represent integers with the same absolute value?

    ```
    V F   T A     P N   U
    -6 -5 -4 -3 -2 -1 0 1 2 3 4 5 6
    ```

    A. points V and U
    B. points F and P
    C. points T and A
    *D. points F and N

2. **SHORT ANSWER** Danielle owes her brother $40. She pays him $25. Write an integer to represent how much she still owes her brother. Explain how you solved.

    -15; $-40 + 25 = -15$

3. How is the fraction $\frac{19}{30}$ written as a decimal?

    F. 0.63
    *G. $0.6\overline{3}$
    H. $0.\overline{63}$
    I. $0.06\overline{3}$

4. Suppose a submarine is diving from the surface of the water at a rate of 80 feet per minute. Which integer represents the depth of the submarine after 7 minutes?

    A. 80
    B. 560
    C. −80
    *D. −560

5. What is the simplified form of the algebraic expression shown below?

    $$7w - 6 - 3w + 5$$

    *F. $4w - 1$
    G. $w + 2$
    H. $w - 1$
    I. $4w - 6$

6. Which expression is equivalent to the algebraic expression below?

    $$-4(3x - 5)$$

    A. $-x - 5$
    B. $-x - 9$
    *C. $-12x + 20$
    D. $-12x - 5$

Benchmark Test Answer Keys

Course 2 Benchmark Test – Second Quarter (continued)

7. Suppose a 24-acre plot of land is being divided into $\frac{1}{3}$-acre lots for a housing development. How many lots will there be in the development?

F. 8 lots
G. 27 lots
H. 56 lots
*I. 72 lots

8. Which property is illustrated by the equation below?

$$\frac{5}{6} \times \frac{6}{5} = 1$$

A. Additive Inverse Property
B. Distributive Property
C. Associative Property of Multiplication
*D. Multiplicative Inverse Property

9. Which of the following shows the rational numbers in order from least to greatest?

F. 58%, $0.6\overline{2}$, $\frac{31}{50}$
G. $0.6\overline{2}$, $\frac{31}{50}$, 58%
*H. 58%, $\frac{31}{50}$, $0.6\overline{2}$
I. $\frac{31}{50}$, 58%, $0.6\overline{2}$

10. SHORT ANSWER Does the pattern below represent an arithmetic sequence? Explain your reasoning.

yes; There are 5 squares in the first figure. Each additional figure has 4 more squares than the previous figure. So, it is arithmetic.

11. Which of the following rational numbers is equivalent to a terminating decimal?

*A. $\frac{17}{20}$
B. $\frac{17}{22}$
C. $\frac{17}{24}$
D. $\frac{17}{26}$

12. Jacob is $5\frac{5}{6}$ feet tall. Linda is $5\frac{1}{4}$ feet tall. How much taller is Jacob?

F. $\frac{1}{3}$ ft
*G. $\frac{7}{12}$ ft
H. $\frac{3}{4}$ ft
I. $1\frac{1}{9}$ ft

Course 2 Benchmark Test – Second Quarter (continued)

13. SHORT ANSWER The table shows Elizabeth's scores for 9 holes of golf. Add the numbers in the middle column to find her total score for 9 holes. Add the integers in the third column to find her total score relative to par.

Hole	Score	Relative to Par
1	4	0
2	5	+1
3	3	0
4	4	0
5	7	+2
6	5	+1
7	4	0
8	5	+1
9	2	−1
Totals	?	?

39; +4

14. Which of the following linear expressions cannot be factored?

A. $15x - 10$
B. $4x + 8$
*C. $3x + 8$
D. $2x - 2$

15. The thickness of a CD is about $\frac{1}{20}$ inch. If Carrie has a stack of 52 CDs, what is the height of the stack?

*F. $2\frac{3}{5}$ in.
G. $2\frac{1}{2}$ in.
H. $\frac{5}{13}$ in.
I. $1\frac{1}{10}$ in.

16. What is the next number in the pattern?

2,916, −972, 324, −108, 36, ...

A. −18
*B. −12
C. 12
D. 18

17. Which of the following number sentences represents the model?

F. $\frac{2}{5} \times \frac{2}{3} = \frac{4}{15}$
G. $\frac{3}{4} \times \frac{1}{3} = \frac{1}{4}$
H. $\frac{2}{3} \times \frac{1}{5} = \frac{2}{15}$
*I. $\frac{2}{5} \times \frac{1}{3} = \frac{2}{15}$

18. What is the quotient of the division problem?

$$\frac{-44}{4}$$

*A. −11
B. −4
C. 4
D. 11

Benchmark Test Answer Keys

Course 2 Benchmark Test – Second Quarter (continued)

19. Which of the following represents the expression below simplified?

    $(4x - 1) + (-6x + 3)$

    *F. $-2x + 2$
    G. $-2x + 3$
    H. $-2x - 1$
    I. $3x - 3$

20. Angela painted $\frac{3}{8}$ of a room. Todd painted $\frac{2}{5}$ of the same room. What part of the room has been painted?

    A. $\frac{1}{40}$
    B. $\frac{5}{13}$
    *C. $\frac{31}{40}$
    D. $\frac{15}{16}$

21. What is the result when the expression $(6x - 3)$ is subtracted from $(-3x + 2)$?

    F. $9x - 5$
    *G. $-9x + 5$
    H. $3x - 1$
    I. $-3x + 1$

Course 2 Benchmark Test – Second Quarter (continued)

22. The models below represent the portion of a pizza that Reggie and Edgar have each eaten.

    How much more of the pizza has Edgar eaten than Reggie?

    A. $\frac{2}{3}$
    B. $\frac{1}{4}$
    *C. $\frac{1}{6}$
    D. $\frac{1}{12}$

23. **SHORT ANSWER** James is using properties of real numbers to prove that $3(-1) = -3$. Identify the missing properties from his proof.

Statements	Properties
$3(0) = 0$	Multiplicative Property of Zero
$3[(-1) + 1] = 0$	a.
$3(-1) + 3(1) = 0$	b.
$3(-1) + 3 = 0$	c.
$3(-1) = -3$	d.

    a. **Additive Inverse Property;**
    b. **Distributive Property;**
    c. **Multiplicative Identity;**
    d. **Subtraction Property of Equality**

24. **SHORT ANSWER** Write the next three terms of the arithmetic sequence below.

    $1, 9, 17, 25, 33, \ldots$

    41, 49, 57

25. Overnight the low temperature dropped to -6 degrees Fahrenheit. If the high temperature during the day was 11 degrees Fahrenheit, what was the difference between the high and low temperatures?

    F. $5°F$
    *G. $17°F$
    H. $-5°F$
    I. $-17°F$

Benchmark Test Answer Keys

Course 2 Benchmark Test – Third Quarter

1. Suppose the length of each side of a square is increased by 5 feet. If the perimeter of the square is now 56 feet, what were the original side lengths of the square?

   *A. 9 ft
   B. 11 ft
   C. 14 ft
   D. 36 ft

2. Which operation should be performed first to solve the inequality below?

   $-3x + 5 \leq 23$

   F. add 5 to each side
   G. divide each side by -3
   H. reverse the inequality symbol
   *I. subtract 5 from each side

3. What is the measure of x in the figure below?

   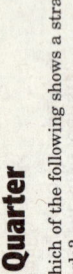

   A. 25°
   B. 65°
   *C. 115°
   D. 125°

4. Which of the following shows a straight angle?

   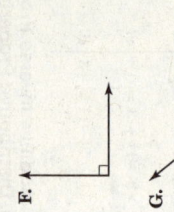

5. What is the solution to the equation below?

   $\dfrac{x}{3} = -6$

   *A. -18
   B. -9
   C. -3
   D. -2

6. **SHORT ANSWER** The sum of the measures of the angles of a triangle is 180°. Write and solve an equation to find the missing measure in the figure below. Show your work.

   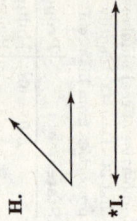

   85; $60 + 35 + n = 180$

Course 2 Benchmark Test – Third Quarter (continued)

7. Which of the following best classifies the triangle below by its angles and sides?

   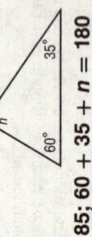

   F. acute, isosceles
   G. acute, equilateral
   H. right, scalene
   *I. right, isosceles

8. **SHORT ANSWER** A shipping company charges $3.50 plus $0.85 per pound to ship a package. Janet shipped a package and the total charge was $8.60. Write and solve an equation to find the weight of the package.

   6 pounds; $3.50 + 0.85n = 8.60$

9. Which of the following describes the shape resulting from the cross section below?

   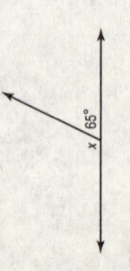

   *A. circle
   B. oval
   C. rectangle
   D. line

10. What is the solution to the equation below?

    $-\dfrac{5}{4}x + \dfrac{2}{5} = -\dfrac{13}{30}$

    F. $\dfrac{2}{75}$
    G. $\dfrac{25}{24}$
    *H. $\dfrac{2}{3}$
    I. $\dfrac{11}{12}$

11. Terrance is making a scale model of a car that is 16 feet long. He is using the scale 1 inch = 2.5 feet. How long is Terrance's model?

    A. 5.8 in.
    *B. 6.4 in.
    C. 28 in.
    D. 40 in.

12. Which of the following describes the shape resulting from the cross section below?

    *F. rectangle
    G. square
    H. triangle
    I. parallelogram

Benchmark Test Answer Keys

Course 2 Benchmark Test – Third Quarter (continued)

13. SHORT ANSWER Carla and Mandy are solving the inequality below.

$$-4x \geq 12$$

Carla says the solution is $x \leq -3$, while Mandy says the solution is $x \geq -3$. Which student is correct? What mistake was made by the other student?

Carla; Sample answer: Mandy forgot to reverse the inequality symbol.

14. Which number line shows the solution to the inequality below?

$$v - 2 > 1$$

A. 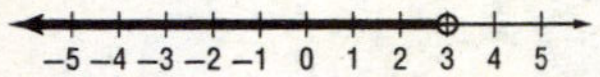
B. 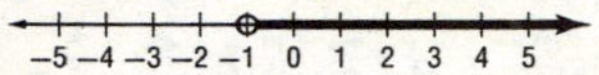
*C. 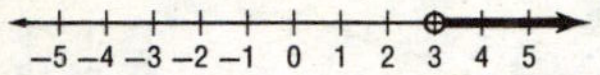
D.

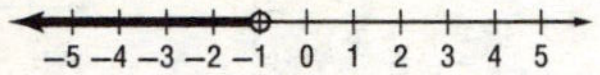

15. Angles R and Z are complementary. If $m\angle R = 26°$, what is the measure of angle Z?

F. 26°
*G. 64°
H. 74°
I. 154°

16. Tien bought movie tickets for herself and two of her friends. She paid $8.50 for each ticket. If Tien has $14.50 left, how much money did she have before she bought the movie tickets?

A. $28.00
B. $31.50
C. $37.50
*D. $40.00

17. The angle measures of a triangle are 28°, 70°, and 82°. Which of the following best classifies the triangle by its angle measures?

*F. acute
G. obtuse
H. right
I. scalene

18. What type of angle is shown below?

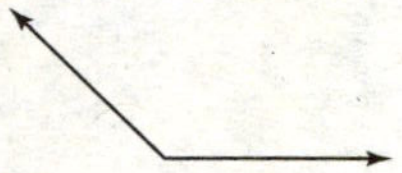

A. acute
B. right
*C. obtuse
D. straight

Course 2 Benchmark Test – Third Quarter (continued)

19. Which of the following is a possible cross section of the figure below?

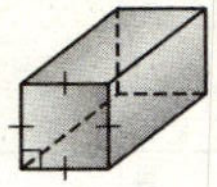

F. triangle
G. hexagon
*H. square
I. trapezoid

20. What is the scale factor of a drawing if the scale is 1 inch = 6 feet?

*A. $\frac{1}{72}$
B. $\frac{1}{6}$
C. 6
D. 72

21. Fran wants to rent a scooter for the afternoon, but she can spend no more than $50.

Scooter Rental	
First Hour	$12.50
Each Additional Hour	$7.50

Which inequality can Fran use to find the maximum number of hours she can rent a scooter?

*F. $12.5 + 7.5n \leq 50$
G. $12.5 + 7.5n < 50$
H. $12.5n + 7.5 \leq 50$
I. $20n < 50$

22. SHORT ANSWER Solve the equation below. Check your answer.

$$2(x + 5) = 16$$

3

23. Five more than twice a number is equal to 19. What is the number?

A. 6
*B. 7
C. 12
D. 28

24. Two angle measures in a parallelogram are labeled. Which term best describes the angles?

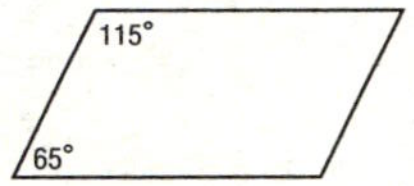

F. complementary
G. acute
H. obtuse
*I. supplementary

Benchmark Test Answer Keys

Course 2 Benchmark Test – End of Year

1. If Michelle rollerblades around a circular track with a radius of 80 meters, how far does she skate? Use 3.14 for π. Round to the nearest tenth.

A. 251.2 m
*B. 502.4 m
C. 12,352 m
D. 20,096 m

2. A sprinter runs 400 meters in 54 seconds. What is the runner's average running rate in meters per second? Round to the nearest tenth.

F. 8.5 meters per second
G. 7.8 meters per second
*H. 7.4 meters per second
I. 6.8 meters per second

3. SHORT ANSWER Find the slope of the line that passes through points A and B. Show your work.

$-\dfrac{3}{4}$

4. The weight of an object on Mars varies directly as the weight of the object on Earth. A 90-pound object on Earth weighs 34 pounds on Mars. If an object weighs 135 pounds on Earth, how much does it weigh on Mars?

*A. 51 pounds
B. 63 pounds
C. 219 pounds
D. 357 pounds

5. A jar contains 3 pennies, 5 nickels, 4 dimes, and 6 quarters. If a coin is selected at random, what is the probability of selecting a penny?

F. $\dfrac{5}{18}$
G. $\dfrac{2}{9}$
H. $\dfrac{1}{3}$
*I. $\dfrac{1}{6}$

6. Which expression is equivalent to the algebraic expression below?

$3(-2x - 1)$

A. $x + 2$
B. $x - 1$
*C. $-6x - 3$
D. $-6x - 1$

Course 2 Benchmark Test – Third Quarter *(continued)*

25. SHORT ANSWER Jamal built the three-dimensional figure below using blocks.

Sketch the front, side, and top views of the figure.

Benchmark Test Answer Keys

Course 2 Benchmark Test – End of Year

7. SHORT ANSWER A cereal company is giving away 1 of 6 different prizes in each box of cereal. Describe a simulation you could use to estimate the number of boxes needed to get all 6 prizes.

Sample answer: Roll a number cube. Let each number represent a different prize. Count the number of rolls needed to get all 6 numbers. Repeat several times and take an average.

8. Which three-dimensional figure is modeled by the net below?

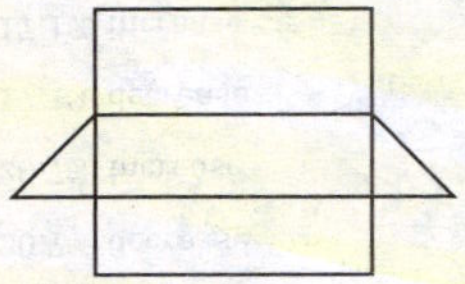

F. rectangular prism
*G. triangular prism
H. square pyramid
I. rectangular pyramid

9. What is the vertical cross section of a cylinder?

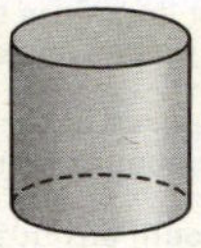

A. circle
B. oval
*C. rectangle
D. point

10. What is the probability of tossing a penny and landing on heads three times in a row?

F. $\frac{3}{2}$
G. $\frac{1}{2}$
H. $\frac{1}{4}$
*I. $\frac{1}{8}$

11. What type of angle is shown below?

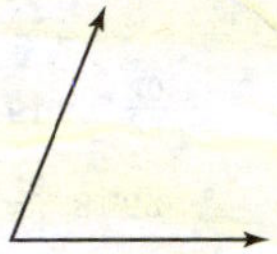

*A. acute
B. right
C. obtuse
D. straight

12. What is the scale factor of a drawing if the scale is 1 inch = 4 feet?

*F. $\frac{1}{48}$
G. $\frac{1}{4}$
H. 4
I. 48

Course 2 Benchmark Test – End of Year (continued)

13. Megan surveyed a random sample of 60 students at her school and found that 42 of them ride the bus to school each day. If there are 320 students at Megan's school, about how many of them ride the bus to school each day?

A. 348 students
*B. 224 students
C. 188 students
D. 132 students

14. SHORT ANSWER The advertisement below shows the terms of a certificate of deposit (CD) at a local bank.

Super CD!
▸ Invest for 2 years and earn 2.75% simple annual interest.
▸ Invest for 3 years and earn 3.25% simple annual interest.
▸ Invest for 4 years and earn 3.75% simple annual interest.

See an associate today!

Suppose Robert invests $1,200 in the CD for a period of 3 years. How much interest will he earn? How much will Robert have after 3 years?

$117; $1,317

15. Last summer there were 88 players at Coach Rodriguez's basketball camp. This year there are 125% of this number of players. How many players are there at camp this year?

F. 70 players
G. 98 players
H. 106 players
*I. 110 players

16. What is the volume of the pyramid shown below?

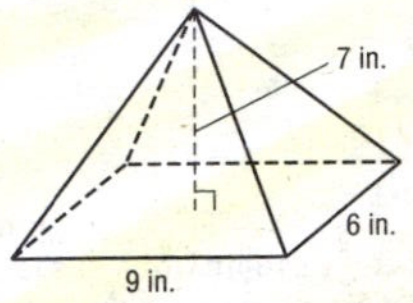

*A. 126 in³
B. 189 in³
C. 221 in³
D. 378 in³

17. What is the constant rate of change of the ordered pairs?

x	y
1	2
3	10
5	18
7	26

F. 8
G. 6
*H. 4
I. 2

18. What is the decimal equivalent of the fraction $\frac{32}{45}$?

A. 0.71
*B. $0.7\overline{1}$
C. $0.\overline{71}$
D. $0.07\overline{1}$

Benchmark Test Answer Keys

Course 2 Benchmark Test – End of Year (continued)

19. Kyle wants to determine the most popular sport among students at his school. Which of the following will likely result in a biased sample?

F. surveying every 5th student standing in the lunch line

G. surveying a random sample of 3 students from each homeroom

*H. surveying a random sample of 25 students attending a school football game

I. surveying every 10th student who enters the school one morning

20. Last year there were 29 students at a creative writing workshop. This year 35 students attended the workshop. To the nearest tenth, what is the percent of change in the number of students in attendance?

A. 20.7% decrease
*B. 20.7% increase
C. 17.1% decrease
D. 17.1% increase

21. In a recent survey, 88% of shoppers at a grocery store said they would be interested in a rewards program. If there were 450 shoppers surveyed, which proportion can be used to find the number who are interested in a rewards program?

F. $\frac{100}{88} = \frac{n}{450}$
G. $\frac{88}{450} = \frac{n}{100}$
H. $\frac{88}{100} = \frac{450}{n}$
*I. $\frac{88}{100} = \frac{n}{450}$

Course 2 Benchmark Test – End of Year (continued)

22. Which property is illustrated by the equation below?

$$7 + (-7) = 0$$

*A. Additive Inverse Property
B. Distributive Property
C. Associative Property of Addition
D. Additive Identity Property

23. Which of the following shows the rational numbers in order from least to greatest?

*F. $81.5\%, 0.81\overline{5}, \frac{33}{40}$
G. $81.5\%, \frac{33}{40}, 0.81\overline{5}$
H. $0.81\overline{5}, \frac{33}{40}, 81.5\%$
I. $0.81\overline{5}, 81.5\%, \frac{33}{40}$

24. SHORT ANSWER The line graph shows the performance of a stock over a 5-day period. Describe what is misleading about the data display.

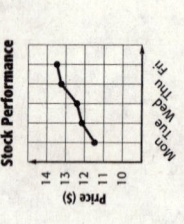

Sample answer: The vertical scale goes from 0 to 10 and then by ones. This makes the performance looks better.

Course 2 Benchmark Test – End of Year (continued)

25. How many blocks were needed to make the rectangular prism below?

A. 54 blocks
*B. 72 blocks
C. 84 blocks
D. 108 blocks

26. Which of the following angles would be classified as an acute angle?

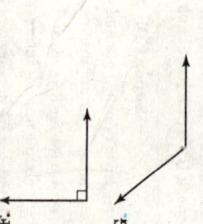

F.
G.
*H.
I.

27. SHORT ANSWER Ronaldo rolled a number cube 50 times. During these trials he rolled the number 5 a total of 7 times. Based on these trials, what is the probability of rolling a 5? Does this represent a theoretical or experimental probability? Explain.

0.14; experimental; it is based on actual experimental results.

28. Which of the following linear expressions *cannot* be factored?

*A. $15x + 22$
B. $12x - 10$
C. $8x - 2$
D. $7x + 21$

29. Which of the following number sentences represent the model shown below?

F. $\frac{3}{4} \times \frac{1}{8} = \frac{3}{32}$
G. $\frac{3}{8} \times \frac{3}{4} = \frac{9}{32}$
*H. $\frac{3}{8} \times \frac{1}{4} = \frac{3}{32}$
I. $1\frac{1}{4} \times 1\frac{1}{3} = \frac{1}{12}$

30. Which of the following rational numbers is equivalent to a repeating decimal?

A. $\frac{24}{60}$
B. $\frac{30}{64}$
C. $\frac{29}{50}$
*D. $\frac{35}{60}$

Course 2 Benchmark Test – End of Year (continued)

31. The angle measures of a triangle are 33°, 94°, and 53°. Which of the following best classifies the triangle by its angle measures?

F. acute
*G. obtuse
H. right
I. scalene

32. SHORT ANSWER Write and solve an equation to find the missing measure. Show your work.

$35; 55 + 90 + n = 180$

33. What is the measure of x in the figure below?

A. 31°
*B. 41°
C. 49°
D. 131°

34. A large pizza at Angelo's Pizzeria has a diameter of 14 inches. What is the area of the pizza? Use 3.14 for π. Round to the nearest tenth.

F. 44.0 in²
G. 122.7 in²
*H. 153.9 in²
I. 615.4 in²

35. A home improvement store normally sells 20-foot extension ladders for $225. This week the ladders are discounted by 20%. What is the sale price of the ladders?

*A. $180
B. $165
C. $60
D. $45

36. SHORT ANSWER A computer store builds custom computers by allowing customers to choose 1 of 4 different CPUs, 1 of 8 hard drives, and 1 of 3 video cards. How many different computers are possible?

96 computers

Course 2 Benchmark Test – End of Year (continued)

37. Which of the following best classifies the triangle below by its angles and sides?

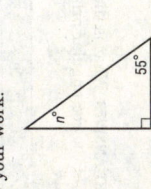

F. acute, isosceles
*G. acute, equilateral
H. acute, scalene
I. obtuse, equilateral

38. In an obstacle course race, how many ways can five finalists be ordered?

A. 1
B. 5
C. 24
*D. 120

39. SHORT ANSWER Compare and contrast the data represented in the double box plot below.

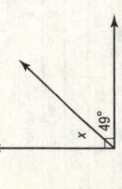

Sample answer: The data from each group have the same median and upper and lower extremes, but the data for group A is clustered more closely around the median.

40. What is the solution to the equation below?

$$\frac{7}{8}\left(x - \frac{1}{2}\right) = -\frac{49}{80}$$

F. $-\frac{6}{5}$
*G. $-\frac{1}{5}$
H. $\frac{1}{5}$
I. $\frac{6}{5}$

41. The table shows the number of yards jogged by Kaylee each minute.

Time (min)	Distance (yd)
1	175
2	350
3	525
4	700

If the pattern continues, how many yards will Kaylee have jogged after 20 minutes?

A. 875 yd
B. 1,750 yd
*C. 3,500 yd
D. 3,850 yd

42. Simplify the expression below.

$$(-7x + 4) - (2x - 8)$$

F. $-5x - 4$
G. $-5x + 12$
H. $-9x - 4$
*I. $-9x + 12$

Benchmark Test Answer Keys

Course 2 Benchmark Test – End of Year (continued)

43. The table shows the number of different types of rides at an amusement park. Which type of data display would be best show the number of items in specific categories?

Type of Ride	Number
Water Slides	9
Rollercoasters	14
Spinning Rides	5
Funhouses	4

*A. bar graph
B. circle graph
C. line graph
D. line plot

44. SHORT ANSWER What is the surface area of the rectangular prism shown below?

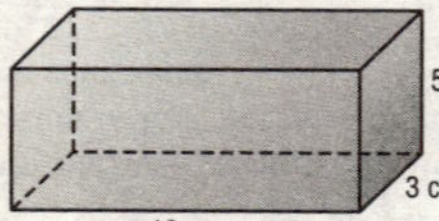

222 cm²

45. Angles C and E are supplementary. If $m\angle C = 77°$, what is the measure of angle E?

F. 13°
G. 77°
*H. 103°
I. 113°

46. How much simple interest would be earned on an investment of $16,000 if the money is invested for 20 years at an annual interest rate of 5.25%?

A. $840
*B. $16,800
C. $16,840
D. $32,800

47. A muffin recipe calls for 8 cups of flour and yields 24 muffins. If Natalie wants to make 60 muffins, how much flour will she need?

F. 180 cups
G. 24 cups
*H. 20 cups
I. 3.2 cups

48. Which number line shows the solution to the inequality below?

$$-4g < 4$$

A. 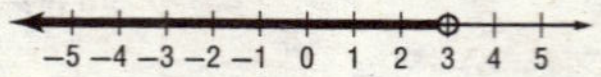
*B. 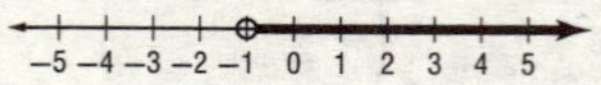
C. 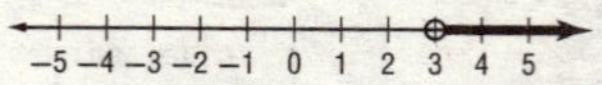
D.

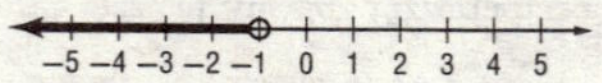

Course 2 Benchmark Test – End of Year (continued)

49. What is the area of the figure below? Use 3.14 for π. Round to the nearest tenth.

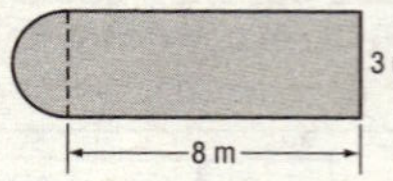

F. 24.0 m²
*G. 27.5 m²
H. 31.1 m²
I. 38.1 m²

50. Christy drove 132 miles in $2\frac{3}{4}$ hours. What was her average speed in miles per hour?

*A. 48 miles per hour
B. 46 miles per hour
C. 44 miles per hour
D. 42 miles per hour

51. The square pyramid has base side lengths of 12 centimeters and a slant height of 15 centimeters. What is the total surface area of the pyramid?

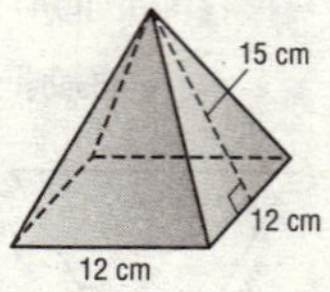

F. 720 in²
G. 640 in²
*H. 504 in²
I. 360 in²

52. Suppose the length of each side of a square is decreased by 4 feet. If the perimeter of the square is now 32 feet, what was the original length of each side?

A. 48 ft
B. 44 ft
C. 16 ft
*D. 12 ft

53. Which of the following is the simplest form of the algebraic expression shown below?

$$-11g + 5 + 6g - 2$$

F. $-6g + 4$
G. $-6g + 5$
H. $-5g + 5$
*I. $-5g + 3$

54. SHORT ANSWER Jamal built the three-dimensional figure below using blocks.

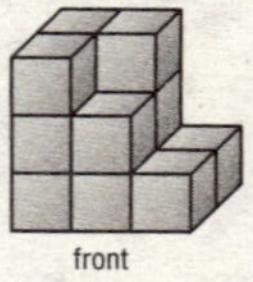

Sketch the front, side, and top views of the figure.

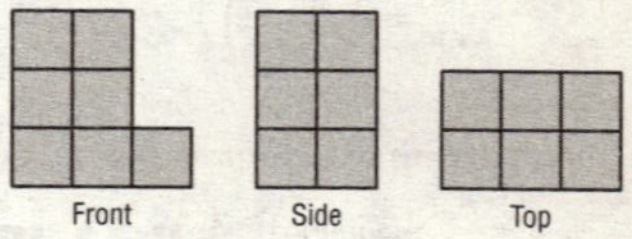

Course 2 Benchmark Test – End of Year (continued)

55. Which of the following represents two dependent events?

   *A. drawing a card from a deck, not replacing it, and drawing another card
   B. rolling a number cube and flipping a coin
   C. drawing a card from a deck, replacing it, and drawing another card
   D. rolling two numbers cubes

56. Which of the following is a possible cross section of a rectangular prism?

   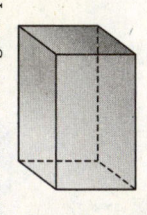

   *F. rectangle
   G. oval
   H. triangle
   I. trapezoid

57. What is the solution to the equation?
   $$4(x + 1) = -16$$

   A. −3
   *B. −5
   C. −63
   D. −65

58. **SHORT ANSWER** Find the volume and surface area of the composite figure shown below if the figure is built with unit cubes.

   volume: 18 cubic units; surface area: 46 square units

59. Which operation should be performed last to solve the inequality below?
   $$-7x + 4 > -10$$

   F. add 4 to each side
   G. subtract 4 from each side
   H. multiply each side by −7 and reverse the inequality symbol
   *I. divide each side by −7 and reverse the inequality symbol

60. What is the product of the expression?
   $$-7(-3)$$

   A. −21
   B. −10
   C. 10
   *D. 21